Reliability Technology Theory and Its Application for Manufacturing Industries

Reliability Technology Theory and Its Application for Manufacturing Industries

Shakuntla Singla
Arun Kumar

ELIVA PRESS

ISBN: 978-99949-8-339-1

Cover Design: Eliva Press
Cover Image: Ingimage
Email: info@elivapress.com
Website: www.elivapress.com

This Eliva Press imprint is published by the registered company Eliva Press Global Ltd. part of Eliva Press S.R.L. Publishing Group

The registered company address is: Pope Hennessy Street Level 2, Hennessy Tower Port Louis, Mauritius
Eliva Press S.R.L. Publishing Group address is: Bulevardul Moscova 21, Chisinau, Moldova, Europe

RELIABILITY TECHNOLOGY THEORY AND ITS APPLICATION FOR MANUFACTURING INDUSTRIES

RELIABILITY TECHNOLOGY THEORY AND ITS APPLICATION FOR MANUFACTURING INDUSTRIES

Dr. Shakuntla Singla[1], Dr. Arun Kumar[2]

[1]Associate Professor, Department of Mathematics and Humanities, MMEC, Maharishi Markandeshwar (Deemed to be University), Mullana, Ambala, India

[2]Assistant Professor, Department of Mathematics, Shri Khushal Das University Hanumangarh, Rajasthan, India

PREFACE

The study of operational research started during the Second World War and afterwards. With the development of operational research, the study of reliability theory emerged as by product in context of defense studies. The words reliable and reliability are in use from ancient time. In fact these occur frequently in social, political, economic and practical fields to indicate the efficiency of a person or mechanical equipment. A mathematical shape to the word reliability was given later in 1950 with its scientific use for defense purpose. Realizing its importance, the study of reliability theory was developed in western world. The development of reliability technology in India is an interesting and encouraging history for researchers. The theory of reliability plays an important role, directly or indirectly in almost all of our daily life problems. Some of the systems whose reliability is of immediate concern to the society in general are power, transportation, medical care, steel and communication industries etc. The history of modern engineering reflects that system failures can occur in any field. Industrial accident in Union Carbide, Bhopal in 1984 and power reactor accident in Chernobyl, USSR in 1986 are prime examples of complex system failure. The reliability analysis of an industry can help the management in taking timely decision for its smooth functioning. This can also help the management to understand the effects of increasing/decreasing repair rate of a particular component or sub system. In order, to obtain maximum output it is necessary to run each of the unit in good condition, i.e. each part of the equipment of the unit should run failure free. Therefore, in the present analysis we have focused on the work about reliability, modeling and analysis of some process industrial systems.

Dr. Shakuntla Singla, Dr. Arun Kumar

TABLE OF CONTENTS

CHAPTER-I

INTRODUCTION

The reliability of a system, equipment and a product is very important aspect of quality for its consistence performance over its expected life span. In fact, uninterrupted service and hazard free operation is an essential requirement of large complex systems like electric power generation and distribution plants or communication network systems such as railways, airways etc. In these cases, a sudden failure of even a single component, assembly or system results in health hazard, accident or interruption in continuity of service.

Thermal power plants provide electric power for domestic, commercial, industrial and agricultural use. Reliability problems may reduce generation of power resulting in load shedding and many other problems including loss of productive activities. Failure of any one system of an aircraft may result in forced landing or an accident. Sudden stoppage of suburban railway train due to fault in carriage system, interruption in the power supply or faulty track sets up a chain of events leading to disruption of services or accidents.

Similarly, sudden failure of a car brake system while it is running may cause serious accident. Causes which are true for the failure of power plants, aircrafts, railways etc. are also true for other products like washing machine, mixer grinder etc. Although the failure of such products may cause inconvenience on smaller scale. The problem of assuring and maintaining reliability has many responsible factors such as original equipment design, control of quality during manufacturing, acceptance inspection, field trials, life testing and design modifications. Therefore, deficiencies in design and manufacturing of such complex systems need to be detected by elaborate testing at the development stage and later corrected by a planned program of maintenance.

This section of the present chapter is introductory in nature. In section 1.1 we first discuss the industrial significance of determining the reliability and availability of the process industries. The basic concepts and definition of reliability are extensively presented in section 1.2. Section 1.3 deals with the renewal theory of reliability. In section 1.4 we have discussed how to estimate the reliability and other parameters to improve the performance of the systems. The methodology for availability analysis using Markova model as well as supplementary variable technique has been discussed in section 1.5. A brief survey of the literature available on the subject is presented in section 1.6. The summary of the work presented in the subsequent chapters of this thesis and objectives of the present work are finally presented in section 1.7.

1.1 INDUSTRIAL SIGNIFICANCE OF STUDYING RELIABILITY AND AVAILABILITY OF PROCESS INDUSTRIES

The reliability of complex systems has emerged as a thrust area because of occurrence of disastrous event in the industries. The manufacturers are highly concerned about reliability of the systems. The manufacturers lose billions of dollars every year as a direct consequence of the unreliability of industrial plants in terms of cost of lost production, cost of fixing or replacing equipment and of course the loss of human life that cannot be measured in terms of money. Huge amount is spent annually on the planned maintenance of industrial systems and assets in order to maintain certain levels of reliability. So, reliability has become one of the vital ingredients in system planning, design, development and operational phase of the system. Its study can benefit industries in terms of higher productivity and low maintenance costs. The events such as the loss of space shuttle Columbia, the chemical spills at Bhopal (India) and recent electricity outage in North America are prime examples of complex system failures.

To compete with the global market and to achieve higher production goals, the industrial system should remain operative for maximum possible duration. Actually, these systems are subjected to random failures. These failures may be due to poor designs, wrong manufacturing techniques, lack of operative skills and experience, adoption of poor maintenance policies, power fluctuations, operations at overload/under load, delay in starting the maintenance, delay in getting the equipment's behavior information, organizational rigidity and complexity and many a times human error also. Therefore, to compete in the global market, high production and good quality (operation and performance wise) is must and can be achieved by maintaining system failure at the lowest possible level (i.e. highest system availability).

Concept of system reliability was developed during the last six decades mainly in late 1940's and early 1950's. The fields of communication and transportation were perhaps the first to witness the rapid growth in complexity due to the advancement in electronics and control systems. Reliability engineering has been stressed for several years in the field of military-aircraft manufacturers. In aircraft design, avionics subsystem was more complex and hence been used for reliability analysis. In 1965, a complex missile program ran into hundreds of millions of dollars which included a reliability program. In this program was a line of high reliability, premium quality and costly parts, designed to be used with sizable safety factors. Another area of modern technology involving reliability was the space program. It was observed and realized that the percentage of successful space launchings has increased dramatically since the early days of our space program due to the application of reliability concept. Transportation sector is other area where the reliability has found major success. Reliability has now become a matter to concern of all those who are working in the industries.

The process industries comprise of large complex engineering system and sub systems arranged in series, parallel or in combination of both. For efficient and economical operation of process plant, each system and sub system should run failure free for long duration under the existing operative conditions. Therefore, all the activities concerning the utilization of manpower, machines, materials and supporting resources must be well organized and coordinated to develop strategies for the optimal utilization. It increases production and hence profit of the concerned industry.

In recent years, with the increase of automation and need for greater cost-effectiveness, the process industries have increasingly become aware that maintenance should no longer be considered simply as supporting service to production process but must be regarded as an integral part of it. From the literature available, it has been found that no rigid system for maintenance can be applied universally to process industries to accommodate every situation. Therefore a suitable maintenance system must be designed and developed to suit the requirements of a particular process industry. A detailed behavior analysis and scientific maintenance planning helps the equipment's/ systems to remain available for longer time. In order to express the system availability in quantitative terms, it is necessary to develop mathematical models for the system/ sub systems and analyze their behaviors to evaluate the performance in real operating conditions.

1.2 BASIC CONCEPTS AND DEFINITIONS

Reliability of a system (product) deals with the concept of dependability, successful operation or performance and the absence of failures. Reliability is the probability that a machine (product) can perform its intended function, without failure, for a specified interval of time when operating under standard conditions. It should be observed that the above stated definition stresses four significant factors: probability, intended function, time and operating conditions. These four elements play an important role in characterizing the reliability of an item. The concept of reliability has been interpreted in number of different ways, out of which few are listed below:-

i) Reliability is the probability that the device operate without failures for a given time under the specified operating conditions.

ii) Reliability of a system is failure free operation for a definite period, under the given operating conditions with minimum time lost for repair and preventive maintenance.

iii) The reliability of an equipment is assumed to be the capacity of the equipment to maintain given properties under specified conditions for a period of time.

Let N_0 be the size of the population out of which N_s units survive the test while N_f fail, then reliability function $R\ (t)$ is given by

$$R(t) = \frac{N_S}{N_o} = \frac{N_o - N_f}{N_o} \tag{1.1}$$

$$\frac{dR(t)}{dt} = -\frac{1}{N_o}\frac{dN_f}{dt} \qquad \text{(Taking } N_0 \text{ fixed)} \tag{1.2}$$

The rate at which component fails can be defined as

$$\frac{dN_f}{dt} = -N_o \frac{dR(t)}{dt} \tag{1.3}$$

Dividing both sides of equation (1.3) by N_s, we obtain the instantaneous probability $r(t)$of failure , that is,

$$r(t) = \frac{1}{N_S}\frac{dN_f}{dt} = -\frac{N_o}{N_S}\frac{dR(t)}{dt} \tag{1.4}$$

Using (1.1) in (1.4), we get

$$r(t) = -\frac{1}{R(t)}\frac{dR(t)}{dt} \tag{1.5}$$

Integrating (1.5), we get

$$\int r(t)dt = -logR(t) \tag{1.6}$$

$$R(t) = \exp[-\int_0^t r(t)dt] \tag{1.7}$$

The function $r(t)$ is called the failure rate function or hazard function or hazard rate.

Equation (1.7) can be considered as a generic expression of failure as it is applicable to both exponential and non-exponential failure distributions.

Failure Rate (λ): The failure rate is expressed in terms of failures per unit. It is computed as the ratio of number of failures of the items undergoing the test time.

$$\lambda = \frac{N_f}{T}, \tag{1.8}$$

where λ= failure rate, N_f = No. of failures during test interval, T = Total test time.

Repair rate (μ): The repair rate is expressed in terms of repairs per unit time. It is computed as the ratio of number of repairs of the items undergoing the test time.

$$\mu = \frac{N_r}{T}, \tag{1.9}$$

where μ= repair rate, N_r = No. of repair during test interval T = Total test time.

Mean Time between Failures (MTBF): Mean time between failures is referred as the average time of satisfactory operation of the system. During the operating period, when failure rate is fairly constant, the MTBF is the reciprocal of the constant failure rate or the ratio of the test time to the number of failures.

$$m \;=\; 1/\lambda \;=\; T/f \tag{1.10}$$

Mean Time to Failure (MTTF): If the failure time of the life test information of n items is $t_1, t_2, t_3, \ldots, t_n$, then the mean time to failure (MTTF) is defined as:

$$MTTF = \frac{1}{n}\sum_{i=1}^{n} t_i \quad (1.11)$$

1.2.1 Availability: Availability is a performance criterion for repairable systems that accounts for both the reliability and maintainability aspects of a system. It is defined as the probability that the system is operating properly when it is required for use. That is, availability is the probability that a system is not in the failed state or undergoing a repair action when it needs to be used. The numerical value of availability is expressed as a probability from 0 to 1. Availability calculations take into accounts both the failures and repairs of the system. For example, if a lamp has 99.9% availability, then there will be one time out of a thousand that someone needs to use the lamp but it is non-operational because of the switch is broken, or it is waiting for the replacement of lamps etc.

Availability Classification: The definition of availability is somewhat flexible, depending on what types of downtimes are considered in the analysis. As a result, there are a number of different classifications of availability:

(i) Point (instantaneous) Availability **(ii)** Average Up-Time Availability (Mean Availability)

(iii) Steady State Availability **(iv)** Operational Availability

Point (instantaneous) Availability($A(t)$): Point or instantaneous availability is the probability that a system (or component) will be operational at any random time t. It is defined by expected up time of the system.

Average Up-Time (Mean) Availability($A_m(T)$): The mean availability is the proportion of time during a mission or time-period when the system is available for use. It represents the mean value of the instantaneous availability function over the period (0, T):

$$A_m(T) = \frac{1}{T}\int_0^T A(t)dt \quad (1.12)$$

Steady State Availability$A(\infty)$: The steady state availability of the system is the limit of the instantaneous availability function as time approaches infinity.

$$A(\infty) = \lim_{t \to \infty} A(t) \quad (1.13)$$

Operational Availability(A_0): Operational availability is a measure of availability, which includes all experienced sources of downtime. The equation for operational availability is:

$$A_0 = \text{Uptime}/(\text{Operation Cycle}) \quad (1.14)$$

where the operation cycle is the overall time period of operation being investigated, and uptime is the total time the system was functioning during the operating cycle. Thus, operational availability is the availability that the customer actually experiences. It is essentially the a posteriori availability based on actual events that happened to the system.

1.2.2 Maintainability: Maintainability, like reliability, has its own unique and diversified elements. Maintainability is a characteristic of a design, installation, and operation, usually expressed as the probability that a machine can be retained in, or restored to specified operable condition within a specified interval of time when maintenance is required. In other words maintainability measures the ease and speed with which a system can be restored to operational status after a failure occurs. This refers to the aspects of a product that increases its serviceability and reparability, increases the cost effectiveness of maintenance, and ensures that the product meets the requirements for its intended use. Maintainability is also a probability in the same way as reliability, its value lies between zero and one. Good maintainability will ensure that reliable equipment will be available to the users.

Maintainability v/s Reliability: Reliability and maintainability jointly affect the availability of the equipment. Highly reliable equipment or a system may fail rarely, but, if its maintainability is poor, then it takes very long time to repair and decommission once it fails. Thus, the availability of highly reliable equipment may reduce considerably, if the maintenance is poor. Similarly equipment may have very good maintainability, but if it has poor reliability then it would fail frequently and in turn availability would get reduced. Maintainability may be given less importance in some applications like missiles and rocket propulsion etc. but, for general industrial equipments and components, maintainability has to be given more considerations.

Maintenance: Maintenance is defined as any action that restores failed units to an operational condition or retains non-failed units in an operational condition. In earlier days very few terms were used in maintenance management like repair, overhauling, preventive maintenance etc. With the involvement of experts in maintenance management several new terms were invented such as planned, scheduled, routine, periodic, breakdown, corrective and predictive maintenance etc. Maintenance actions can be classified on the basis of planning and criticality/essentiality of the jobs.

a) Breakdown Maintenance: In a breakdown maintenance system, repair is undertaken only after the failure of the equipment. The equipment is allowed to run undisturbed till it fails. However, such a breakdown maintenance system can't work in a big industry having large number of equipments. This type of maintenance is not particularly used in chemical or process industries where the reliability requirement is high and the failure may lead to safety or pollution hazard. Breakdown maintenance in such big process industries proves to be very costly as downtime and restart costs are very high.

b) Corrective Maintenance: Corrective maintenance is the action taken so that a failed system is restored to operational status. This usually involves replacing or repairing a component or subsystem that is responsible for the failure of the whole system. Such maintenance is performed at unpredictable intervals, since a component failure time is not known a priori. The objective of corrective maintenance is to restore the system to satisfactory operative state within the shortest possible time. Corrective maintenance is basically carried out in three steps:

I: Diagnosis of the Problem: The maintenance technician must take time to locate the failed parts or otherwise satisfactorily assess the cause of the system failure.

II: Repair and Replacement of Faulty Components: Once the cause of system failure has been determined, action must be taken to address the cause, usually by replacing or repairing the components that caused the system to fail. Repair action should also be taken in case the component is working below acceptable reduced capacity.

III: Verification of the Repair Action: Once the components have been repaired or replaced, the maintenance technician must verify that the system is operating successfully.

c) Preventive Maintenance (PM): Preventive maintenance is the practice of replacing components or subsystems before they fail in order to promote continuous system operation. Preventive maintenance is also defined as the planned maintenance of plants and equipment in order to prevent or minimize the breakdowns and the depreciation rates. As it covers vast areas, occasionally some people get misled about its coverage. Some people think PM is just a routine inspection, cleaning, lubrication, adjustment and doing minor repairs/jobs on equipment. Some other think that PM means internal cleaning of equipments and components, lubrication and oil changing and replacement of consumables like belts, seals, bearing etc. Yet some other think that PM includes only major jobs like overhauling and reconditioning etc. Actually PM includes all three types of activities mentioned here. After PM repairs, the equipment's health is restored back nearly to the original condition. The schedule for preventive maintenance is based on observation of past behavior of the system. The objective of preventive maintenance is to run the equipments in good conditions for a long time. Cost is also a factor for the scheduling of preventive maintenance. In many circumstances, it is financially more sensible to replace parts or components that have not failed at predetermined intervals rather than to wait for a system failure which may result in costly disruption in operations.

1.3 RENEWAL THEORY

For a repairable system, the time of operation is not continuous. In other words, the system's life cycle can be described by a sequence of up and down states. The system operates until it fails, then it is repaired and returned to its original operating state. It will fail again after some random time of operation, get repaired again and this process of failure and repair will repeat. This is called a renewal process and is defined as a sequence of independent and non-negative random variables. In this case, the random variables are the times-to-failure and the time-to-repair. Each time a unit fails and is restored to working order, a renewal is said to have occurred. This type of renewal process is known as an alternating renewal process, since the state of the component alternates between a functioning state and a repair state.

System Reliability: System reliability is a measure of the performance of the system under the specified conditions. In most of the complex systems it has been observed that, they consist of components and subsystems connected in series, parallel or standby or a combination of these. To calculate system reliability following basic steps are required:

(i) The components and subsystems, which constitute a given system and whose individual reliability factors can be estimated, are identified and computed.

(ii) The configuration in which the components are connected to form the system is represented in logical manner either by a block diagram or by a transition diagram.

(iii) The condition for successful operation of the system is then established, that is, it is decided as how the components should function. For example, we can consider whether all components be operative or it is sufficient that k out of n components function.

(iv) The combination rules of probability theory are stated to be applied to estimate the system reliability.

System reliability can be enhanced by using various techniques as given below:

(i) Parts improvement method

(ii) Effective and creative design

(iii) System simplification

(iv) Structural redundancy

(v) Maintenance and repair

1.4 TYPE OF SYSTEMS

On the basis of repair point of view, the systems can be classified as:

1. Non-repairable system. 2. Repairable system.

1. Non-Repairable System

This type of system operates only once. Such systems have an instantaneous life requirement e.g. fuses, missiles, flash bulbs. Reliability is the important criteria to calculate the effectiveness of non-repairable system.

2. Repairable System:

A) Continuously operating System: This type of system once put in operation continues to operate till its failure or the system is stopped for planned maintenance. Examples are nuclear furnaces, earth satellites etc.

b) Once on and off operating system: This type of system is characterized by the fact that it can be operated and re-operated when desired e.g. turbines, pumps, computer, etc.

c) Intermittently operating system: In this case, the system is always in operational readiness, but is required to operate intermittently e.g. telephone, radar, etc.

On the basis of arrangement of components in the system, the system can be classified as:

1. Series System 2. Parallel System 3. Series-parallel System 4. Stand-by System

Series System: This type of system generally consists of a large number of components connected in series. If anyone of these components fails, the system fails. This is also one of the most commonly used structures.

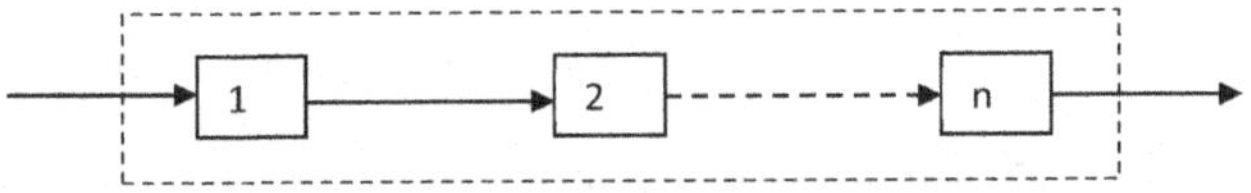

Fig. 1.1: Series combination of components

Let the successful operation of these individual units be represented by X_1, $X_2, \dots, X_n$, and their respective probabilities by P(X_1), P(X_2), ..., P(X_n). For the successful operation of the system, it is necessary that all n units function satisfactorily. Hence the reliability of the n-component series system is given by (assuming that these units are independent)

$$P(s) = P(X_1) \times P(X_2) \times \dots \times P(X_n) = \sum_{i=1}^{n} P(X_i) \quad (1.15)$$

whereas $P(X_i)$ is the reliability of the ith component in series in the system and P(S) is the reliability of the system having n components in series.

Parallel System: Generally this type of system consists of a large number of components connected in parallel. A parallel system is fail only when its all components are fail. Parallel configuration is often referred to as redundancy. Redundancy is the technique which involves more than one component to achieve a higher reliability.

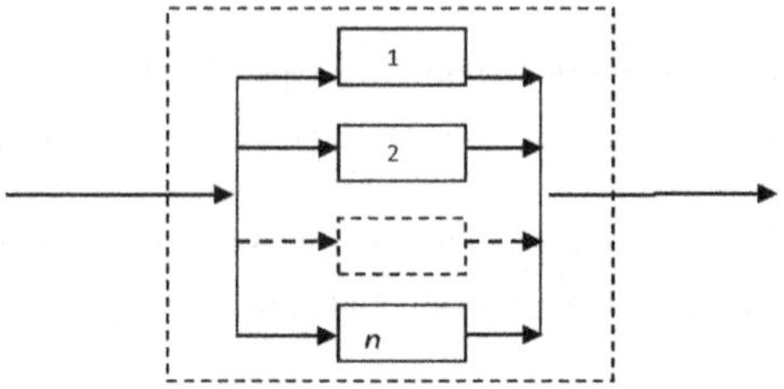

Fig. 1.2: Parallel combination of components

Let the successful operation of these individual units be represented by $X_1, X_2, \dots, X_n$, let $\overline{X_1}, \overline{(X_2)}, \dots, \overline{(X_n)}$, respectively, represent their unsuccessful operation, i.e., the failure of the n units.

$$P(\bar{S}) = P(\overline{X_1}) \times P(\overline{X_2}) \times \dots \times P(\overline{X_n}) = 1 - P(S)$$

Hence $P(S) = 1 - ([1 - P(X_1)] \times [1 - P(X_2)] \dots \times [1 - P(X_n)])$ (1.16)

Series-Parallel System: This system consists of stage 1, stage 2,..., stage k connected in series.

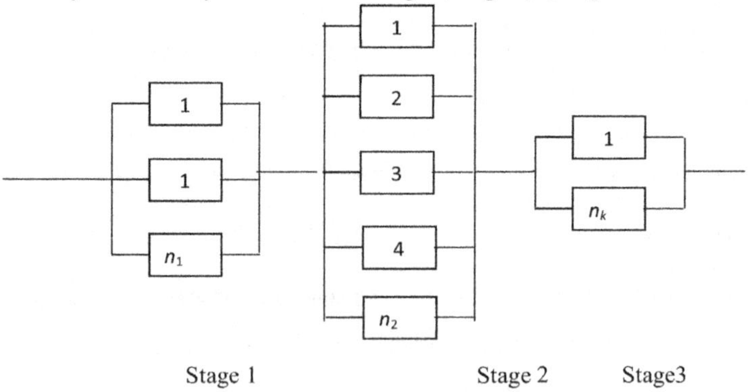

Fig. 1.3: Series-Parallel combination of components

Each stage contains a number of redundant elements, stage 1 consisting of n_i redundant elements connected in parallel. The reliability of such a system is the product of the reliabilities of each stage. Stage i with n_i elements will have the reliability given by

$$R_i = 1 - [1 - P(X_{i1})][1 - P(X_{i2})] \dots [1 - P(X_{in})] = 1 - \sum_{j=1}^{n_i}[1 - P(X_{ij})]$$

$$R(s) = R_1 R_2 R_3 \dots R_K = \sum_{i=1}^{k}\left\{1 - \sum_{j=1}^{n_i}[1 - P(X_{ij})]\right\} \quad (1.17)$$

Redundancy: If the state of the art is such that either it is not possible to produce highly reliable components or the cost of producing such components is very high, then we can improve the reliability of the system by introducing the technique of redundancies. This involves the deliberate creation of new parallel paths in a system. There are many methods of introducing redundancies in a system of which few are mentioned below:

a) **Element Redundancy:** In element redundancy two elements are connected in parallel. In this, all the channels or paths are active from the beginning of the system till its failure. In this case the reliability of the system is higher than individual reliabilities.

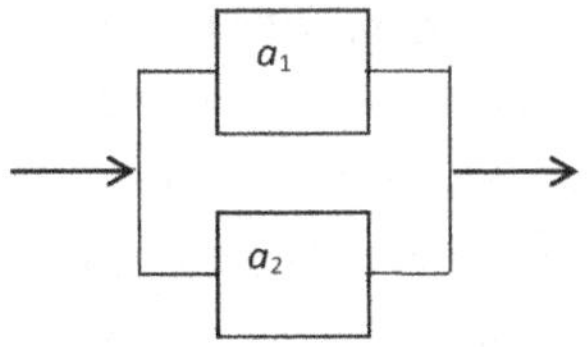

Fig. 1.4: Element Redundancy

b) Standby Redundancy: This is another type of redundancy that can be introduced in a system. In this type of configuration, all the paths are not active at the same time. Initially, when the system is put on operation, the ideal switch or decision switch connects the input end to element a_1 while the other element a_2 is kept in reserve in a turned-off condition. If elements a_1 fails due some to reason, the decision switch senses it by an in-built mechanism and the connection is made to the standby element a_2.

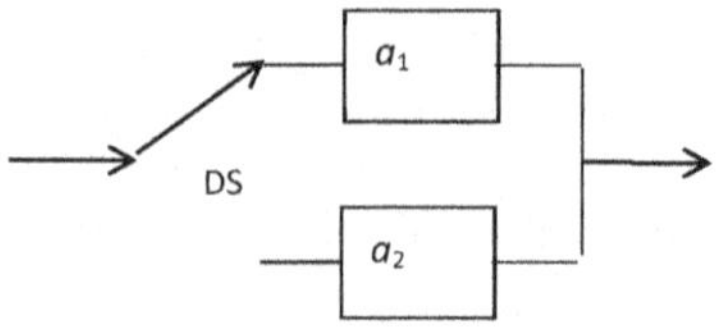

Fig. 1.5 Standby Redundancy

1.5 AVAILABILITY ANALYSIS METHOD USING MARKOV MODEL AND SUPPLEMENTARY VARIABLE TECHNIQUE

In order to find the availability of a system one has to form a system of linear differential equations using mnemonic rule. This rule states that the derivative of the probability of every state is equal to the sum of all probability flows which come from other states to the given state minus the sum of all probability flows which go out from the given state to the other states. The differential equations thus derived are known as the Chapman-Kolmogorov differential equations.

The calculation of the availability of a system with elements exhibiting dependent failures and involving repair or standby operation is, in general, complicated and several approaches have been suggested to carry out the computations. In case the failure and repair rates are variable then it loses its markovian property. By Introducing supplementary variables, the non marlovian character of the system is changed to makovian. Any Markov model is defined by a set of probabilities P_{ij} which define the probability of transition from any state $'i'$ to another$'j'$. This transition probability depends only on states $'i'$and $'j'$, and is independent of all previous states except the last one, i.e., state $'i'$.

The failure rate $\lambda(y)$ and the repair rate $\beta(x)$ are variable. In order to illustrate the availability of the system, let us consider a system consisting of a single repairable component 'A' with a variable failure and repair rate $\lambda(y)$ and $\beta(x)$, respectively.

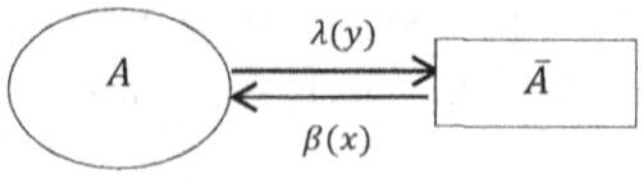

Fig. 1.6: Transition diagram of two components

For this system, the possible states are:

(i) State $A(t)$, where the system is in state A at time t

(ii) State $\bar{A}(x, y, t)$, where the system is in state $\bar{A}$ at time t and has an elapsed failure time 'y' and elapsed repair time 'x'.

If the system remains in state A there can be two possibilities as:

(i) That system is in state 'A ' at time t and no failure occurs in the interval$(t,\ t + \Delta t)$. Probability of such a state is given by $P_A\ (t)(1 - \lambda(y)\Delta t)$.

(ii) That system is in state '$\bar{A}$' at time t and the repair is carried out in the interval $(t, t\ +\ \Delta t)$. Probability of such a state is given by $P_{\bar{A}}(x, y, t)\ (\beta(x)\Delta t)$. Following the mnemonic rule and transition diagram (fig.1.6). The resulting equation for the system in state A can be written as:

$$P_A\ (t + \Delta t) = \ (I - \lambda\ (y)\Delta t)\ P_A\ (t)\ +\ \int [(\beta(x).\Delta t). P_{\bar{A}}(x, y, t)]dx\, dy$$

$$\frac{P_A(t + \Delta t) - P_A\ (t)}{\Delta t} = -\lambda(y)P_A(t) + \int [(\beta(x))\ P_{\bar{A}}(x, y, t)]dx\, dy$$

taking limit as $\Delta t\ \rightarrow\ 0$, we have

$$\frac{d}{dt}P_A(t) = -\lambda(y)P_A(t) + H_0(t) \tag{1.18}$$

where $H_0(t) = \int [(\beta(x).\Delta t). P_{\bar{A}}(x, y, t)]dx\, dy$

Similarly, when the system is in state '$\bar{A}(x, y, t)$' the equation will be

$$\frac{P_{\bar{A}}(x, y, t + \Delta t) - P_{\bar{A}}\ (x, y, t)}{\Delta t} + \frac{P_{\bar{A}}(x + \Delta x, y, t) - P_{\bar{A}}\ (x, y, t)}{\Delta x} + \frac{P_{\bar{A}}(x, y + \Delta y, t) - P_{\bar{A}}\ (x, y, t)}{\Delta y} = -\beta(x)\ P_{\bar{A}}(x, y, t) +\ \lambda\ (y)\ P_A(t))$$

taking limit as $\Delta t\ \rightarrow\ 0, \Delta x \rightarrow 0, \Delta y \rightarrow 0$, we have

$$[\frac{\partial}{\partial t} + \frac{\partial}{\partial x} + \frac{\partial}{\partial y}]P_{\bar{A}}(x, y, t)\ =\ -\beta(x)\ P_{\bar{A}}(x, y, t) +\ \lambda\ (y)\ P_A(t)) \tag{1.19}$$

Initial Conditions: As elapsed failure and repair time are zero initially and the system is completely in working state, the initial conditions thus becomes:

$$P_{\bar{A}}(x, y, 0) = 0; \quad (1.20)$$

$$P_A(0) = 1 \quad (1.21)$$

Boundary condition: Since a system is in the failed state with failure rate $\lambda(y)$ but repair has not been done at that time, so the boundary condition is:

$$P_{\bar{A}}(0, y, t) = \lambda(y) P_A(t) \quad (1.22)$$

SOLUTION OF EQUATIONS

Equation (1.18) is a first order ordinary differential equation and equation (1.19) is a linear partial differential equation which constitutes a Chapman Kolmogorov differential equations. In order to find the reliability of the system, the governing equations (1.18-1.19) will be solved along with the initial and boundary conditions. Equation (1.18) is ordinary differential equation, so can be integrated directly whereas the equation (1.19) is a partial differential equation which can be solved by using Lagrange's method and thus we have

$$P_A(t) = e^{-\int \lambda(y) dt}\left[1 + \int e^{-\int \lambda(y) dt} H_0(t)\ dt\right] \quad (1.23)$$

$$P_{\bar{A}}(x, y, t) = e^{-\int \beta(x) dx}[\lambda(y - x) P_A(t - x) + \int \lambda(y) P_A(t) e^{\int \beta(x) dx} dx] \quad (1.24)$$

The time dependent Availability $A(t)$ of the system is next computed as:

$$A(t) = P_A(t) \quad (1.25)$$

Special Case: When all the transition rates, that is failure and repair are constant then $\frac{\partial}{\partial x}, \frac{\partial}{\partial y} \rightarrow 0$. Consequently equation (1.18-1.19) reduce to the ordinary linear differential equation:

$$\frac{d}{dt} P_A(t) = -\lambda P_A(t) + H_0(t) \quad (1.26)$$

$$[\frac{d}{dt} + \beta] P_{\bar{A}}(t) \ = \ \lambda P_A(t) \quad (1.27)$$

The differential equation (1.26) – (1.27) can be solved analytically by using Laplace transformation method, matrix calculus or numerically following the approach of Gupta (2003) by using the initial conditions:

$$P_A(t) = 1 \text{ and } P_{\bar{A}}(t) = 0 \qquad (1.28)$$

1.6 LITERATURE SURVEY

The world reliable and reliability are used in social, political, business and technical field for expressing a faith/trust in a person, firm or equipment. The reliability analysis of an industry can help the management in taking timely decision for its smooth functioning. In 1960, first text book on reliability by Dummer and Griffen appeared in literature. Since then a number of research papers have been published in the field of reliability (1960, 1963, 1964, 1969, 1978). Professor R. S. Verma and his students Mohan, Garg and Sangal (1962) at Delhi University initiated the study of reliability technology. Singh (1976) first time used reliability technology to analyse the working of production system. Dhillon and Singh (1981) discussed the basic theory of reliability in their book entitled "Engineering reliability-new technique and applications". The availability of working systems used in daily life was calculated by Singh and Aggarwal (1979). He along with his co-workers studied the behavior of process industries. Yang and Dhillon (1996) calculated the availability of a robot with safety system. The need and application of reliability technology in the process industry was discussed by Michelsen (1998). sKumar *et al.* (1989, 1990, and 1992) calculated the availability for number of systems in process industries. Gupta *et al.* (2005) discussed the reliability and availability analysis of serial processes of butter oil plant and behavior analysis of the cement industry. Agnihotri *et al.* (2008) have studied the reliability analysis of boiler used in readymade garment industry.

It has also been observed that environmental conditions may affect the system's reliability considerably. Such systems are very much peculiar in nature. Reliability of mechanical component under environment stress was discussed by Dhillon (1980). Singh (1980) discussed the reliability analysis under fluctuating environment. Aggarwal and Kumar (1986) discussed the stochastic behavior of repairable system working under fluctuating weather conditions. Mokadies *et al.* (2010) discussed the comparison between two cold and warm standby outdoor electric power systems in changing weather.

To improve the reliability, the concept of redundancy is introduced in the system. Kapoor and Kapoor (1976) discussed the effect of standby redundancy on the system's reliability. Gupta and Sharma (1987) evaluated the reliability behavior of a complex system consisting of three subsystems connected in parallel redundancy using Boolean function technique. Sharma (1974) calculated the time dependent of a complex standby redundant system under preemptive repeat repair discipline. Nakagava and Osaki (1975) considered stochastic behavior of a two-unit priority standby redundant system. Aven (1990) discussed availability for standby systems of identical units. Singh and Goel (1996) discussed the availability of heating system with warm standby and imperfect switch in sugar industry. The standby redundancy allocation in series and parallel systems was discussed by Misra *et al.* (2011).

The reliability models with different constraints of operation and repair on standby units were analyzed by Tuteja, Taneja and Vashistha (2001). The reliability and profit analysis of two-unit standby system was discussed by Tuteja and Taneja (1991, 1992, 1993). Lie and Jian (2011) considered repairable systems with repairman, who can take multiple vacations. If the system fails and the repair man is on vacation then system will wait for repair until the repair man is available. The steady state availability of a repairable system consisting of one operating and one spare unit was studied by Jau and Yunn (2007).

A system can fail due to common errors and human errors. Dhillon (1977) discussed a A4 unit redundant system with common cause failure. Later on Singh (1989) calculated the reliability of a warm standby system with common cause failures. The important factor of critical human error, which causes complete failure of the system, was introduced by Dhillon and Misra (1984). Sharma *et al.* (1985) suggested a methodology for prevention of failure and repair policy for systems with common cause failures. Gupta and Kumar (1987) evaluated the availability and mean time to failure of a two-unit cold standby system with three possible states of units, that is, good, partially failed and failed state by introducing the concept of human repair. Chung (1987, 1989) extended the idea to a repairable system subjected to failure due to common cause failure and critical human error. Such systems are very common in our day to day life. Kaushik and Singh (1994) performed the reliability analysis of the naphtha fuel oil and water system under priority repair used in thermal power plant. Sridharan and Mohanavadivu (1996) discussed the reliability and availability analysis for two identical unit parallel systems with common cause failure and human errors.

The priority in repair is given to the components which work at 100% capacity rather than standby component which work at 30% capacity. Kuo-Hsiung *et al.* (2006) discussed the four different systems with warm standby components and standby switching failures based on their reliability and availability.

k-out-of-n structure is also a very popular type of redundancy and is applied in industrial and military systems. Reliability and availability of such systems have been analyzed by various researchers including Chiang and Niu (1981), Chaung (1990), Shao and Lamberson (1991), Li and Chen (2004). Pham (2010) evaluated the modeling of a shared load k-out of-n: g system. The analysis of consecutive k-out of -n: f systems with single repair facility were discussed by Kumar and Gopalan (1997). Vanderperre (2004) discussed the reliability analysis of a renewable multiple cold standby system. The effect of switch failure on 2-redudant system was discussed by Singh (1980).

Cost is the most important factor to increase the availability of the process industries. Prabhuswami (1997) studied the reliability based optimization of manufacturing systems. Gupta *et al.* (1993) discussed with profit analysis of a two unit priority standby system subject to degradation and random shocks. Subramanian and Anantharaman (1994) carried out reliability analysis of a complex standby redundant system, and estimated the comprehensive cost function. Profit analysis of two unit cold standby system was discussed by Siwach *et al.* (2001). Zhao (1994) discussed the availability for repairable components and series systems. The dependability modeling using petri net based model was discussed by Malhotra and Trivedi (1995). Vanderperre (2000) calculated the long run availability of a two unit standby system subjected to a priority rule. Chander and Bansal (2005) discussed the profit analysis of single unit reliability models at different failure modes. Wang and Chiu (2007) calculated the cost benefit of the availability of a warm standby unit with imperfect coverage. The cost analysis of two dissimilar units was discussed by Mokadies and Matta (2010).

Maintainability and availability are two main aspects, which are closely related to reliability. In a reliable system, breakdowns are less frequent and hence availability is high (i.e. system functions well and is available for use). Maintenance analysis helps in determining how often the system and its components should be maintained for reliable performance. Over the last two decades, many methods/techniques have been developed / presented in number of research papers to determine the optimal maintenance schedule. Barlow and Hunter (1960) studied the preventive maintenance models with minimum repairs. Gaver (1963) suggested a method to estimate maintenance performance. Fukuta and Kodama (1974) discussed the mission reliability for a redundant repairable system with two dissimilar units. Nakagawa (1977) developed a model for imperfect preventive maintenance in which the effective age of the system is reduced by 'x' the time units at each preventive maintenance. In (1980) he also developed optimum preventive maintenance policy for repairable system. Gandhi and Wani (1999) evaluated maintainability index of mechanical systems using digraph and matrix method. An algorithm for preventive maintenance policy was developed by Lie and Chaun (1986). Ntuen (1991) proposed a generalized models for determining minimum cost preventive maintenance. Jayabalan and Chaudhary (1992) presented a model for cost optimization of maintenance scheduling for a system with assured reliability. The reliability optimization of complex systems through-SOMGA was studied by Kusumdeep and Dipti (2009). A multi objective optimization of imperfective preventive maintenance policy with hidden failure rates was calculated by Wang and Pham (2011).

Dekker (1995) has given an overview on the role of operation research model for taking maintenance decision. Schabe (1995) presented a method for obtaining optimum replacement time of a complex system. Dijkuizen and Heijden (1995) proposed a series of mathematical models and optimization techniques, to obtain the optimal preventive maintenance. Vaurio *et al.* (1999) discussed the availability and cost functions for periodically inspected preventively maintained units. Ma *et al.* (2001) calculated the optimization of a preventive maintenance scheduling for semiconductor manufacturing systems. Chan and Asgarpoor (2001) discussed the preventive maintenance with markov process. Grall *et al.* (2002) presented a preventive maintenance structure for a gradually deteriorating single-unit system. Ramakrishna and Bawa (2005) have discussed optimization of machine design criteria for higher reliability and maintainability in food processing industry.

The earlier researchers in the field of reliability analyzed the systems using Laplace Transform and matrix method. Most of these workers discussed the systems exhibiting markovian properties. A methodology for failure analysis in process plant was developed by Priel (1974). Singh (1977) discussed the preemptive repeat priority repairs and failure of non-failed component during system failure of complex system. Kumar *et al.* (1991) discussed the behavior analysis of paper production system with different repair policies. Cox (1955) analyzed the non markovian system using supplementary variables. The system having non markovian property can be converted to a system having markovian nature, by introducing some new variables called supplementary variables. Weiss (1962) introduced semi markov process to solve maintainability problem. Singh and Dayal (1989) used supplementary variable technique for problem formulation. Zhang (1996) studied the stochastic behavior of standby system under preemptive priority repair and obtained the expression for transient and steady state of the system using techniques of supplementary variables and Laplace transforms. Khan and Kabir (1995) described a simulation modeling technique for assessing the availability of ammonia plant whereas Gurov and Utkin (1995) discussed the time dependent availability of repairable m-out-of-n and cold standby systems using arbitrary distributions and repair facility. Agnihotri and Satsangi (1996) carried out reliability analysis of a two non- identical unit under priority repair and inspection using regenerative point technique. Perman *et al.* (1997) presented the markovian application to power plant. Hsu (1999) presented a simultaneous determination of preventive maintenance and replacement policies in a queue like production system with minimal repair.

Gyemin and Jongwoo (1999) suggested a new approach for the analysis of a N/G/1 finite queue with the supplementary variable method. Alfa and Rao (2000) formalized the supplementary variable technique for stochastic models. Zhang and Horigome (2001) calculated the availability and reliability of systems with dependent components and time varying failure and repair rates. Habchi (2002) discussed an improved method of reliability assessment for suspended test whereas Gunes and Deveci (2002) have studied the reliability of service systems and its application in student office. Malik *et al.* (2008) performed the reliability and profit evaluation of an operating system with different repair strategy. Kiureghain and Ditlevson (2007) discussed the availability, reliability and downtime of system with repairable components. Gupta (2004) calculated the availability by taking repair rates as variable and failure rates as constant.

1.7 PRESENT WORK

The significant work has been done on the reliability and availability of some industrial systems. In practice, repair and failure rates of industrial systems have been considered to be constant. Only few authors have made an attempt up to a certain extent to study the time dependent case and the case of variable failure and repair rates. Still lot of work needs to be done as far as process industries are concerned.

Reliability and availability of various industrial systems such as Rice, Steel, and Polytube manufacturing plant with constant failure and repair rates have been partly discussed by several authors. In-depth investigations to see how reliability and availability will be affected with variable repair and failure rates are very much in need. Therefore, we intend to address some of these problems in the present study. We propose to investigate the following problems:

To study the reliability and availability of the process industries by considering the time dependent case and the case of variable failure and repair rates. This problem is discussed by making use of supplementary variable technique.

To consider the process industries such as Rice, Steel and Polytube for their reliability modeling and analysis. Since the management is always interested in long run availability, so this case is discussed as special case of these process industries. Other useful parameters like availability and MTBF etc. will also be evaluated.

Keeping these objectives in view, an effort has been made in the present thesis to develop mathematical models to analyze the reliability and availability of some process industries based on actual data taken from the industries. However, manufacturing plants such as Steel, Polytube and Rice plants have been previously discussed by several authors for reliability and availability analysis by assuming constant failure and repair rates. The reliability and availability of these manufacturing plants with variable failure and repair rates have not been discussed satisfactory.

Chapter wise summary of the work presented in the subsequent chapters of this thesis is as follows:

In second chapter, the time dependent and long run availability of polytube manufacturing plant has been analyzed. The plant consists of five sub-systems, namely, mixer, extruder, die, cutter and trifling chute. The failure rate of trifling chute has been found to be very low and assume to have never failed. As per the industry the subsystems cutter and die can work in the reduced state. The transition rates of the all subsystems which lead to the failed state of the system have been assumed to be variable. Behavior analysis of the system has been discussed by assuming that the subsystems cutter and die fail simultaneously. The governing equation determining the reliability of the polytube manufacturing plant has been derived using mnemonic rule from the transition diagram. The equation thus formed is known as Chapman Kolmogorov differential equation consisting of the system of partial differential equations. This partial differential equation has been solved by using Lagrange's method to obtain the time dependent availability. The effect of failure and repair rates of the subsystems on long run availability has been analyzed taking the real data from the process industry.

Chapter III deals with the availability of polytube industry when the subsystems cutter and die fail independently. The governing equations determining the availability of the polytube manufacturing plant has been derived using mnemonic rule from the transition diagram. The Chapman Kolmogrorov differential equations have been thus formed. These partial differential equations are next solved by using Lagrange's method to obtain the time dependent availability. The long run availability has been analyzed by taking constant failure and repair rates of the subsystems. Certain conclusions based on the study have also been presented in this chapter.

In chapter IV the behavioral analysis of time dependent availability and long run availability of rice plant has been discussed. This plant has six units namely: elevator, husking, separator, cleaning, whitening and polishing, all working in series. The concept of pending failed state has been introduced to perform preventive maintenance. This problem has been studied when preventive maintenance of the husk and separator machines are performed. As multi component repairable systems with variable failure and repair rate distribution are difficult to handle mathematically, supplementary variables have been introduced to change the non markovian character into markovian character. The mathematical formulation for determining the availability of rice plant has been carried out in the form of system of partial differential equations. The partial differential equations thus obtained have been solved by using Lagrange's method to obtain the state probabilities for various choices of constant failure and repair rates of the subsystems. We have also considered the constant failure and repair rates of the system. With these assumptions the governing partial differential equations reduce to systems of ordinary differential equations. The system thus obtained has finally been solved numerically by using Runge-Kutta fourth order method for suitable choices of constant failure and repair rates. The long run availability of the plant is also computed by solving the equations recursively. The effect of failure and repair rates on availability of the system has been analyzed critically with the help of tables and diagram.

The analysis of time dependent and long availability of the main parts of the steel industry having mixed, series and parallel configuration has been studied in chapter V. This industry consists of four subsystems. The first subsystem grinding machine is subjected to major failure and the subsystems descaling and the hot steckel machine work in the reduce capacity. The whole system remains operative for a short period of time. The system works under the assumptions that failure and repair rates are variable. The mathematical formulation has next been carried out by using supplementary variable technique. The governing partial differential equations thus formed have been solved by using Lagrange's method for various combinations of variable failure and repair rates. As a special case again failure and repair rates are taken as constants and the governing differential equations have been reduced to system of ordinary differential equations which have further been solved numerically using the fourth order Runge -Kutta method. The effects of failure and repair rates of the subsystems on long run availability of the system have been studied by taking the real time data obtained from the process industry.

Based on the present study of various process industries conclusions have been finally presented in chapter VI. The industrial significance along with the limitations and scope of the present work has also been briefly discussed in this concluding chapter.

CHAPTER-II

AVAILABILITY ANALYSIS OF POLY-TUBE MANUFACTURING PLANT WHEN THE SUBSYSTEMS DIE AND CUTTER FAIL SIMULTANEOUSLY

Polytube provide a sustainable and safe way to distribute the gas and liquid from one place to another. Their environmental contribution to public health and sanitation is unique. Moreover, their long service life followed by recycling, guarantees a high level of sustainability for many years. Plastic pipes are more durable and economical to install, operate and maintain. They have much longer life than iron pipes as iron is easily corrosive in the wet and humid conditions, which decrease its age and make it weaker and prone to leakage. Keeping in view the importance of polytube in real life, we have chosen polytube manufacturing plant in this chapter. This plant is situated near Kurukshetra, India. The industry management provided us all the necessary information about this plant which consisting of five subsystems namely mixture, extruder, die, cutter and trifling chute. While discussing the functioning of various subsystems they mentioned some special features of the subsystems cutter and die. These two subsystems can fail simultaneously as well as independently. As per requirement of the concerned industry, we have proposed the study of availability analysis of the polytube manufacturing plant under the two assumptions: (i) when the subsystems cutter and die fail simultaneously (ii) when the subsystems cutter and die fail independently. Keeping these points in view, the availability analysis of the polytube manufacturing plant, when both subsystems cutter and die fail simultaneously, has been carried out in this chapter and the availability analysis of the same manufacturing plant under the second assumption will be discussed in the succeeding chapter.

This chapter is organized as follows: Section 2.1 is introductory in nature and discusses how polytube is manufactured. Distribution of various subsystems required in the manufacturing of polytube as well as their notation is presented in section 2.2. This section also deals with certain assumptions under which the performance of polytube manufacturing plant depends. Mathematical formulation of Chapman-Kolmogorov differential equations when subsystems are fail simultaneously has been developed in section 2.3.

In this section the transient and steady states have also been presented when the failure and repair rates are variable and constant respectively. The mathematical problem thus developed has next been solved analytically and numerically in section 2.4. The behavior analysis of the system in transient and steady states are discussed in section 2.5. The conclusion based on the performance analysis is finally drawn in section 2.6.

2.1 INTRODUCTION

To manufacture plastic pipe, industry uses a process known as Profile Extrusion. The plastic is extruded over a mandrill which forms it into a tube. This tube is sent down to a cooling apparatus where it is cooled using water, air, or their combination. It is then sent to a cutter where it is cut to various lengths. This process is given in the fig. 2.1.

2.2 SYSTEM DESCRIPTION, NOTATIONS AND ASSUMPTIONS

The process flow chart of the polytube manufacturing plant has been shown in fig. 2.1. The polytube manufacturing plant consists of following five sub-systems.

Sub-system A (Mixture): It mixes raw material such as PVC rising, calcium carbonate, wax and other chemicals in appropriate proportion for manufacturing the pipe. It consists of a heater by which the raw material is heated up to 130^0 C and then transported to the extruder by conveyors. It consists of blades and a motor whose failure causes complete failure of the system.

Sub-system B (Extruder): Raw material obtained from mixture is heated in this section. It consists of a heater to heat the raw material at different temperatures. The quality of the product depends on heating process. Its failure causes the complete failure of the system.

Sub-system C (Die): It is used to make different sizes of pipes. Minor failure of this subsystem reduces the capacity of the system and hence causes loss in production. Major failure of this subsystem results in complete failure of the system.

Sub-system D (Cutter): This sub-system has two units arranged in series. First unit is blade, which cuts the pipe and the second unit is motor which cuts the pipe in different sizes. Failure of blade reduces the capacity of the system while the failure of motor causes the complete failure of the system.

Sub-system E (Trifling Chute): It is used to measure the length of the pipe and is also used for collecting the pipes. This system is supposed to never fail.

In addition to the notations used for subsystems, i.e. .A, B, C, D and F, we have also used the following notations:

A, B, C, D:	Indicate that the subsystems are working in full capacity.
$\bar{C}, \bar{D}$:	Indicate the reduced state of the subsystems C and D.
a, b, c, d :	Indicate the failed state of the subsystems $A, B, C, and\ D$.
$\alpha_i(y)(\mathrm{i} = 1, \dots, 4)$:	Failure rate of the subsystems A, B, C and D respectively.
$\phi(x), \psi(x), \mu(x)$ and $\sigma(x)$:	General repair rates of A, B, C and D respectively.
$P_o(t)$:	Probability that the system is working in full capacity.
$P_i(x, y, t)$ $(i = 1, \dots, 16)$:	Probability that the system is in state i at time t and has an elapsed failure time y and elapsed repair time x.

All the assumptions mentioned below are taken as such as provided by the chosen industry. The assumptions, on which the present analysis is based upon, are as follows:

(i) Repair and failure rates are independent of each other and their unit of measurement is taken as per day.

(ii) Failure and repair rates of the subsystems are taken as variable.

(iii) Performance wise, a repaired unit is as good as new one for a specified duration.

(iv) Sufficient repair facilities are provided.

(v) Service of the subsystem includes repair and/or replacement.

(vi) Subsystems C and D can work in reduced capacity also.

(vii) There are simultaneous failures.

Based on the above notations and assumptions, the transition diagram of the system is given in fig. 2.2.

2.3 MATHEMATICAL FORMULATION OF THE SYSTEM WHEN TWO SUBSYSTEMS CUTTER AND DIE FAIL SIMULTANEOUSELY

Probability considerations give the following system of differential difference equation associated with the state transition diagram (fig. 2.2) of these systems at time $(t + \Delta t)$. We first develop the equation in transient zeroth state by using mnemonic rule, as under:

2.3.1 Transient State when failure and repair rates are variable:

$$
\begin{aligned}
P_0(t+\Delta t) = {} & [1-\alpha_1(y)\Delta t-\alpha_2(y)\Delta t-\alpha_3(y)\Delta t-\alpha_4(y)\Delta t]P_0(t) \\
& + \int \mu(x)P_1(x,y,t)dx\, dy\Delta t \\
& + \int \sigma(x)P_2(x,y,t)dxdy\Delta t + \int \phi(x)P_4(x,y,t)dxdy\, \Delta t \\
& + \int \psi(x)P_5(x,y,t)dxdy\, \Delta t
\end{aligned}
$$

$$
\begin{aligned}
P_0(t+\Delta t) - P_0(t) & \\
= {} & -[\alpha_1(y)\Delta t+\alpha_2(y)\Delta t+\alpha_3(y)\Delta t+\alpha_4(y)\Delta t]P_0(t) \\
& + \int \mu(x)P_1(x,y,t)dxdy\, \Delta t \\
& + \int \sigma(x)P_2(x,y,t)dxdy\Delta t + \int \phi(x)P_4(x,y,t)dx\, \Delta t \\
& + \int \psi(x)P_5(x,y,t)dxdy\, \Delta t
\end{aligned}
$$

dividing both sides by Δt, we get

$$
\begin{aligned}
\frac{P_0(t+\Delta t) - P_0(t)}{\Delta t} & \\
= {} & -[\alpha_1(y)+\alpha_2(y)+\alpha_3(y)+\alpha_4(y)]P_0(t) + \int \mu(x)P_1(x,y,t)dxdy \\
& + \int \sigma(x)P_2(x,y,t)dxdy + \int \phi(x)P_4(x,y,t)dx\, dy \\
& + \int \psi(x)P_5(x,y,t)dxdy
\end{aligned}
$$

and on taking limit as $\Delta t \to 0$, we have

$$\left[\frac{d}{dt} + T_0\right] P_0(t) = C_0 \tag{2.1}$$

Similarly, we can write the differential equation for the other states as:

$$\left[\frac{\partial}{\partial t}+\frac{\partial}{\partial x}+\frac{\partial}{\partial y}+T_1(x,y)\right]P_1(x,y,t)=C_1(x,y,t) \tag{2.2}$$

$$\left[\frac{\partial}{\partial t}+\frac{\partial}{\partial x}+\frac{\partial}{\partial y}+T_2(x,y)\right]P_2(x,y,t)=C_2(x,y,t) \tag{2.3}$$

$$\left[\frac{\partial}{\partial t}+\frac{\partial}{\partial x}+\frac{\partial}{\partial y}+T_3(x,y)\right]P_3(x,y,t)=C_3(x,y,t) \tag{2.4}$$

$$\left[\frac{\partial}{\partial t}+\frac{\partial}{\partial x}+\frac{\partial}{\partial y}+\phi(x)\right]P_4(x,y,t)=\alpha_1(y)P_0(t) \tag{2.5}$$

$$\left[\frac{\partial}{\partial t}+\frac{\partial}{\partial x}+\frac{\partial}{\partial y}+\psi(x)\right]P_5(x,y,t)=\alpha_2(y)P_0(t) \tag{2.6}$$

$$\left[\frac{\partial}{\partial t}+\frac{\partial}{\partial x}+\frac{\partial}{\partial y}+\phi(x)\right]P_6(x,y,t)=\alpha_1(y)P_1(x,y,t) \tag{2.7}$$

$$\left[\frac{\partial}{\partial t}+\frac{\partial}{\partial x}+\frac{\partial}{\partial y}+\psi(x)\right]P_7(x,y,t)=\alpha_2(y)P_1(x,y,t) \tag{2.8}$$

$$\left[\frac{\partial}{\partial t}+\frac{\partial}{\partial x}+\frac{\partial}{\partial y}+\mu(x)\right]P_8(x,y,t)=\alpha_3(y)P_1(x,y,t) \tag{2.9}$$

$$\left[\frac{\partial}{\partial t}+\frac{\partial}{\partial x}+\frac{\partial}{\partial y}+\phi(x)\right]P_9(x,y,t)=\alpha_1(y)P_2(x,y,t) \tag{2.10}$$

$$\left[\frac{\partial}{\partial t}+\frac{\partial}{\partial x}+\frac{\partial}{\partial y}+\psi(x)\right]P_{10}(x,y,t)=\alpha_2(y)P_2(x,y,t) \tag{2.11}$$

$$\left[\frac{\partial}{\partial t}+\frac{\partial}{\partial x}+\frac{\partial}{\partial y}+\sigma(x)\right]P_{11}(x,y,t)=\alpha_4(y)P_2(x,y,t) \tag{2.12}$$

$$\left[\frac{\partial}{\partial t}+\frac{\partial}{\partial x}+\frac{\partial}{\partial y}+\phi(x)\right]P_{12}(x,y,t)=\alpha_1(y)P_3(x,y,t) \tag{2.13}$$

$$\left[\frac{\partial}{\partial t}+\frac{\partial}{\partial x}+\frac{\partial}{\partial y}+\psi(x)\right]P_{13}(x,y,t)=\alpha_2(y)P_3(x,y,t) \tag{2.14}$$

$$\left[\frac{\partial}{\partial t}+\frac{\partial}{\partial x}+\frac{\partial}{\partial y}+\mu(x)\right]P_{14}(x,y,t)=\alpha_3(y)P_3(x,y,t) \tag{2.15}$$

$$\left[\frac{\partial}{\partial t}+\frac{\partial}{\partial x}+\frac{\partial}{\partial y}+\sigma(x)\right]P_{15}(x,y,t)=\alpha_4(y)P_3(x,y,t) \tag{2.16}$$

where,

$$T_0 = \alpha_1(y) + \alpha_2(y) + \alpha_3(y) + \alpha_4(y)$$
$$T_1(x,y) = \mu(x)+\alpha_1(y) + \alpha_2(y) + \alpha_3(y) + \alpha_4(y)$$
$$T_2(x,y) = \sigma(x)+\alpha_1(y) + \alpha_2(y) + \alpha_3(y) + \alpha_4(y)$$
$$T_3(x,y) = \sigma(x)+\mu(x) + \alpha_1(y) + \alpha_2(y) + \alpha_3(y) + \alpha_4(y)$$
$$C_0 = \int \mu(x)P_1(x,y,t)dxdy + \int \sigma(x)P_2(x,y,t)dxdy + \int \phi(x)P_4(x,y,t)dxdy + \int \psi(x)P_5(x,y,t)dxdy$$

$$C_1(x,y,t) = \alpha_3(y)P_0(t) + \phi(x)P_6(x,y,t) + \psi(x)P_7(x,y,t) + \mu(x)P_8(x,y,t) + \sigma(x)P_3(x,y,t)$$

$$C_2(x,y,t) = \alpha_4(y)P_0(t) + \phi(x)P_9(x,y,t) + \psi(x)P_{10}(x,y,t) + \mu(x)P_3(x,y,t) + \sigma(x)P_{11}(x,y,t)$$
$$C_3(x,y,t) = \alpha_4(y)P_1(x,y,t) + \alpha_3(y)P_2(x,y,t) + \phi(x)P_{12}(x,y,t) + \psi(x)P_{13}(x,y,t) + \mu(x)P_{14}(x,y,t) + \sigma(x)P_{15}(x,y,t)$$

Boundary Conditions: As discussed in section 1.3 of chapter one the boundary conditions of the subsystems are specified as:

$$P_1(0,y,t) = \alpha_3(y)P_0(t) \tag{2.17}$$

$$P_2(0,y,t) = \alpha_4(y)P_0(t) \tag{2.18}$$

$$P_3(0,y,t) = \int \alpha_4(y)P_1(x,y,t)\,dx + \int \alpha_3(y)P_2(x,y,t)dx \tag{2.19}$$

$$P_4(0,y,t) = \alpha_1(y)P_0(t) \tag{2.20}$$

$$P_5(0,y,t) = \alpha_2(y)P_0(t) \tag{2.21}$$

$$P_6(0,y,t) = \int \alpha_1(y)P_1(x,y,t)\,dx \tag{2.22}$$

$$P_7(0,y,t) = \int \alpha_2(y)P_1(x,y,t)\,dx \tag{2.23}$$

$$P_8(0,y,t) = \int \alpha_3(y)P_1(x,y,t)\,dx \tag{2.24}$$

$$P_9(0,y,t) = \int \alpha_1(y)P_2(x,y,t)\,dx \tag{2.25}$$

$$P_{10}(0,y,t) = \int \alpha_2(y)P_2(x,y,t)\,dx \tag{2.26}$$

$$P_{11}(0,y,t) = \int \alpha_4(y)P_2(x,y,t)\,dx \quad (2.27)$$

$$P_{12}(0,y,t) = \int \alpha_1(y)P_3(x,y,t)\,dx \quad (2.28)$$

$$P_{13}(0,y,t) = \int \alpha_2(y)P_3(x,y,t)\,dx \quad (2.29)$$

$$P_{14}(0,y,t) = \int \alpha_3(y)P_3(x,y,t)\,dx \quad (2.30)$$

$$P_{15}(0,y,t) = \int \alpha_4(y)P_3(x,y,t)\,dx \quad (2.31)$$

Initial Conditions: The initial conditions of the subsystems are as under:

$$P_i(x,y,0) = 0; \qquad (i = 1\ldots15) \quad (2.32)$$

$$P_0(0) = 1 \quad (2.33)$$

On solving the differential equations (2.1-2.16) together with initial and boundary conditions (2.17-2.33), we can find all the probabilities in terms of $P_0(t)$ and then time dependent availability $A(t)$ can be computed.

As a special case we shall now discuss how to develop Chapman Kolmogorov differential equation in transient as well as steady states when both failure and repair rates are constant.

2.3.2 Transient State when failure and repair rates are constant

In order to find the reliability of the system when both failure and repair rates are constant, system of equations (2.1-2.16) reduces to more simplified form which are given below:

$$\left[\frac{d}{dt} + \alpha_1 + \alpha_2 + \alpha_3 + \alpha_4\right]P_0(t) = \mu P_1(t) + \sigma P_2(t) + \phi P_4(t) + \psi P_5(t) \quad (2.34)$$

$$\left[\frac{d}{dt} + \mu + \alpha_1 + \alpha_2 + \alpha_3 + \alpha_4\right]P_1(t) = \alpha_3 P_0(t) + \phi P_6(t) + \psi P_7(t) + \mu P_8(t) + \sigma P_3(t) \quad (2.35)$$

$$\left[\frac{d}{dt} + \sigma + \alpha_1 + \alpha_2 + \alpha_3 + \alpha_4\right]P_2(t) = \alpha_4 P_0(t) + \phi P_9(t) + \psi P_{10}(t) + \mu P_3(t) + \sigma P_{11} \quad (2.36)$$

$$\left[\frac{d}{dt} + \sigma + \mu + \alpha_1 + \alpha_2 + \alpha_3 + \alpha_4\right] P_3(t) = \alpha_4 P_1(t) + \alpha_3 P_2(t) + \phi P_{12}(t) + \psi P_{13}$$
$$+\mu P_{14}(t) + \sigma P_{15}(t)$$
(2.37)

$$\left[\frac{d}{dt} + \phi\right] P_4(t) = \alpha_1 P_0(t) \quad (2.38)$$

$$\left[\frac{d}{dt} + \psi\right] P_5(t) = \alpha_2 P_0 \quad (2.39)$$

$$\left[\frac{d}{dt} + \phi\right] P_6(t) = \alpha_1 P_1(t) \quad (2.40)$$

$$\left[\frac{d}{dt} + \psi\right] P_7(t) = \alpha_2 P_1(t) \quad (2.41)$$

$$\left[\frac{d}{dt} + \mu\right] P_8(t) = \alpha_3 P_1(t) \quad (2.42)$$

$$\left[\frac{d}{dt} + \phi\right] P_9(t) = \alpha_1 P_2(t) \quad (2.43)$$

$$\left[\frac{d}{dt} + \psi\right] P_{10}(t) = \alpha_2 P_2(t) \quad (2.44)$$

$$\left[\frac{d}{dt} + \sigma\right] P_{11}(t) = \alpha_4 P_2(t) \quad (2.45)$$

$$\left[\frac{d}{dt} + \phi\right] P_{12}(t) = \alpha_1 P_3(t) \quad (2.46)$$

$$\left[\frac{d}{dt} + \psi\right] P_{13}(t) = \alpha_2 P_3(t) \quad (2.47)$$

$$\left[\frac{d}{dt} + \mu\right] P_{14}(t) = \alpha_3 P_3(t) \quad (2.48)$$

$$\left[\frac{d}{dt} + \sigma\right] P_{15}(t) = \alpha_4 P_3(t) \quad (2.49)$$

Initial conditions: The initial conditions of the subsystems are as under:

$$P_i(0) = \begin{cases} 1, & i = 0 \\ 0, & \text{otherwse} \end{cases} \quad (2.50)$$

One can obtain the state probabilities $P_i(i = 1, \ldots, 15)$ by solving the system of differential equations (2.34-2.49) together with initial conditions (2.50).

2.3.3 Steady state when failure and repair rates are constant

Management is always interested in long run availability to achieve their optimal target. This can be obtained mathematically by taking $\frac{d}{dt} \to 0$ as $t \to \infty$ in the system of equations (2.34-2.49) therefore, the system of equations (2.34-2.49) reduces to the following system of linear equations:

$$[\alpha_1 + \alpha_2 + \alpha_3 + \alpha_4]P_0 - \mu P_1 - \sigma P_2 - \phi P_4 - \psi P_5 = 0 \qquad (2.51)$$

$$[\mu + \alpha_1 + \alpha_2 + \alpha_3 + \alpha_4]P_1 - \alpha_3 P_0 - \phi P_6 - \psi P_7 - \mu P_8 - \sigma P_3 = 0 \qquad (2.52)$$

$$[\sigma + \alpha_1 + \alpha_2 + \alpha_3 + \alpha_4]P_2 - \alpha_4 P_0 - \phi P_9 - \psi P_{10} - \mu P_3 - \sigma P_{11} = 0 \qquad (2.53)$$

$$[\sigma + \mu + \alpha_1 + \alpha_2 + \alpha_3 + \alpha_4]P_3 - \alpha_4 P_1 - \alpha_3 P_2 - \phi P_{12} - \psi P_{13} - \mu P_{14} - \sigma P_{15} = 0 \qquad (2.54)$$

$$\phi P_4 - \alpha_1 P_0 = 0 \qquad (2.55)$$

$$\psi P_5 - \alpha_2 P_0 = 0 \qquad (2.56)$$

$$\phi P_6 - \alpha_1 P_1 = 0 \qquad (2.57)$$

$$\psi P_7 - \alpha_2 P_1 = 0 \qquad (2.58)$$

$$\mu P_8 - \alpha_3 P_1 = 0 \qquad (2.59)$$

$$\phi P_9 - \alpha_1 P_2 = 0 \qquad (2.60)$$

$$\psi P_{10} - \alpha_2 P_2 = 0 \qquad (2.61)$$

$$\sigma P_{11} - \alpha_4 P_2 = 0 \qquad (2.62)$$

$$\phi P_{12} - \alpha_1 P_3 = 0 \qquad (2.63)$$

$$\psi P_{13} - \alpha_2 P_3 = 0 \qquad (2.64)$$

$$\mu P_{14} - \alpha_3 P_3 = 0 \qquad (2.65)$$

$$\sigma P_{15} - \alpha_4 P_3 = 0 \qquad (2.66)$$

2.4 SOLUTIONS

The system of differential equations (2.1-2.16) together with the boundary conditions (2.17-2.31) and initial conditions (2.32-2.33) is called Chapman- Kolmogorov differential difference equation. Equation (2.1) is a linear differential equation of first order and other equations (2.2) to (2.16) are linear partial differential equations. In order to find the availability of the system, the governing equations (2.1-2.16) along with the boundary conditions (2.17-2.31) and initial conditions (2.32-2.33) have been solved by using Lagrange's method to get probabilities $P_i(t)$ $(i = 1, \dots, 15)$ for each state:

$$P_0(t) = e^{-T_0 t}[1 + \int C_0(t)e^{T_0 t}dt] \tag{2.67}$$

$$P_1(x, y, t) = e^{-\int T_1(x,y)dx}[\int C_1(x, y, t)e^{\int T_1(x,y)dx}dx + \alpha_3(y - x)P_0(t - x)] \tag{2.68}$$

$$P_2(x, y, t) = e^{-\int T_2(x,y)dx}[\int C_2(x, y, t)e^{\int T_2(x,y)dx}dx + \alpha_4(y - x)P_0(t - x)] \tag{2.69}$$

$$P_3(x, y, t) = e^{-\int T_3(x,y)dx}[\int C_3(x, y, t)e^{\int T_3(x,y)dx}dx + \int \alpha_4(y - x)P_1(x, y - x, t - x)\,dx + \int \alpha_3(y - x)P_2(x, y - x, t - x)] \tag{2.70}$$

$$P_4(x, y, t) = e^{-\int \phi(x)dx}[\alpha_1(y - x)P_0(t - x) + \int \alpha_1(y)P_0(t)e^{\int \phi(x)dx}dx] \tag{2.71}$$

$$P_5(x, y, t) = e^{-\int \psi(x)dx}[\alpha_2(y - x)P_0(t - x) + \int \alpha_2(y)P_0(t)e^{\int \psi(x)dx}dx] \tag{2.72}$$

$$P_6(x, y, t) = e^{-\int \phi(x)dx}[\int \alpha_1(y - x)P_1(x, y - x, t - x)\,dx + \int \alpha_1(y)P_1(x, y, t)e^{\int \phi(x)dx}dx] \tag{2.73}$$

$$P_7(x, y, t) = e^{-\int \psi(x)dx}[\int \alpha_2(y - x)P_1(x, y - x, t - x)\,dx + \int \alpha_2(y)P_1(x, y, t)e^{\int \psi(x)dx}dx] \tag{2.74}$$

$$P_8(x, y, t) = e^{-\int \mu(x)dx}[\int \alpha_3(y - x)P_1(x, y - x, t - x)\,dx + \int \alpha_3(y)P_1(x, y, t)e^{\int \mu(x)dx}dx] \tag{2.75}$$

$$P_9(x, y, t) = e^{-\int \phi(x)dx}[\int \alpha_1(y - x)P_2(x, y - x, t - x)\,dx + \int \alpha_1(y)P_2(x, y, t)e^{\int \phi(x)dx}dx] \tag{2.76}$$

$$P_{10}(x,y,t) = e^{-\int \psi(x)dx}[\int \alpha_2(y-x)P_2(x,y-x,t-x)\,dx + \int \alpha_2(y)P_2(x,y,t)e^{\int \psi(x)dx}dx] \quad (2.77)$$

$$P_{11}(x,y,t) = e^{-\int \sigma(x)dx}[\int \alpha_4(y-x)P_2(x,y-x,t-x)\,dx + \int \alpha_4(y)P_2(x,y,t)e^{\int \sigma(x)dx}dx] \quad (2.78)$$

$$P_{12}(x,y,t) = e^{-\int \phi(x)dx}[\int \alpha_1(y-x)P_3(x,y-x,t-x)\,dx + \int \alpha_1(y)P_3(x,y,t)e^{\int \phi(x)dx}dx] \quad (2.79)$$

$$P_{13}(x,y,t) = e^{-\int \psi(x)dx}[\int \alpha_2(y-x)P_3(x,y-x,t-x)\,dx + \int \alpha_2(y)P_3(x,y,t)e^{\int \psi(x)dx}dx] \quad (2.80)$$

$$P_{14}(x,y,t) = e^{-\int \mu(x)dx}[\int \alpha_3(y-x)P_3(x,y-x,t-x)\,dx + \int \alpha_3(y)P_3(x,y,t)e^{\int \mu(x)dx}dx] \quad (2.81)$$

$$P_{15}(x,y,t) = e^{-\int \sigma(x)dx}[\int \alpha_4(y-x)P_3(x,y-x,t-x)\,dx + \int \alpha_4(y)P_3(x,y,t)e^{\int \sigma(x)dx}dx] \quad (2.82)$$

From the above relations, all the probabilities are obtained in terms of $P_0(t)$ given by equation (2.67). Thus, the time dependent Availability $A(t)$ of the system is given as:

$$A(t) = P_0(t) + \int \sum_{i=1}^{3} P_i(x,y,t)\,dxdy \quad (2.83)$$

In order to solve the system of differential equation (2.34-2.49) together with initial conditions (2.50), most of the authors have used Laplace transformation and matrix method to find the probability of the transient states. As it is difficult to find Laplace inverse of higher degree polynomials, so the complexity increases with the increase in number of equations. To overcome such type of problems, the system of differential equation (2.34-2.49) with initial conditions (2.50) has been solved numerically following the approach used by Gupta *et al.* (2005). The numerically computation have been carried out starting from $t = 0$ to $t = 360$ days by assuming $t = 0.005$ as one day. The availability of polytube has been obtained by taking different combinations of constant failure and repair rates of the subsystems collected from the concerned industry. The availability $A(t)$ of the system can be computed as,

$$A(t) = \sum_{i=0}^{3} P_i(t) \tag{2.84}$$

The availability of the system, as defined in (2.84) has been computed for various values of the repair and failure rates. It may be mentioned here that these combinations are not again exhaustive and we have only considered the main subsystems(A, B, C, D) in the numerical study.

Finally the system of linear equations (2.51-2.66) determining the steady state availability has been solved recursively be expressing all the probabilities in terms of P_0. These are obtained as follows:

$$P_1 = M_1 P_0 \tag{2.85}$$

$$P_2 = M_2 P_0 \tag{2.86}$$

$$P_3 = M_3 P_0 \tag{2.87}$$

$$P_4 = \frac{\alpha_1}{\phi} P_0 \tag{2.88}$$

$$P_5 = \frac{\alpha_2}{\psi} P_0 \tag{2.89}$$

$$P_6 = \frac{\alpha_1}{\phi} P_1 \tag{2.90}$$

$$P_7 = \frac{\alpha_2}{\psi} P_1 \tag{2.91}$$

$$P_8 = \frac{\alpha_3}{\mu} P_1 \tag{2.92}$$

$$P_9 = \frac{\alpha_1}{\phi} P_2 \tag{2.93}$$

$$P_{10} = \frac{\alpha_2}{\psi} P_2 \tag{2.94}$$

$$P_{11} = \frac{\alpha_4}{\sigma} P_2 \tag{2.95}$$

$$P_{12} = \frac{\alpha_1}{\phi} P_3 \tag{2.96}$$

$$P_{13} = \frac{\alpha_2}{\psi} P_3 \tag{2.97}$$

$$P_{14} = \frac{\alpha_3}{\mu} P_3 \tag{2.98}$$

$$P_{15} = \frac{\alpha_4}{\sigma} P_3 \tag{2.99}$$

where

$$M_1 = \frac{\alpha_3}{S_3} + \frac{\alpha_3 \sigma \alpha_4}{S_1 S_2 S_3}, \quad M_2 = \frac{\alpha_4}{S_4} + \frac{\mu \alpha_4 M_1}{S_1 S_2}, \quad M_3 = \frac{\alpha_4}{S_1} M_1 + \frac{\alpha_3}{S_1} M_2$$

$$S_1 = \sigma + \mu, \quad S_2 = \sigma + \alpha_3 - \frac{\mu \alpha_3}{S_1}, \quad S_2 = \mu + \alpha_4 - \frac{\sigma \alpha_4}{S_1} - \frac{\mu \alpha_3 \sigma \alpha_4}{S_1 S_2 S_1}$$

Now using the normalizing conditions $\sum_{i=0}^{15} P_i = 1$, we get

$$P_0 = \left[(1 + \frac{\alpha_1}{\phi} + \frac{\alpha_2}{\psi} + \left(1 + \frac{\alpha_1}{\phi} + \frac{\alpha_2}{\psi} + \frac{\alpha_3}{\mu}\right) M_1 + \left(1 + \frac{\alpha_1}{\phi} + \frac{\alpha_2}{\psi} + \frac{\alpha_4}{\sigma}\right) M_2 + (1 + \frac{\alpha_1}{\phi} + \frac{\alpha_2}{\psi} + \frac{\alpha_3}{\mu} + \frac{\alpha_4}{\sigma}) M_3\right]^{-1}$$

(2.100)

Thus, steady state availability of the manufacturing plant is obtained as

$$A(\infty) = \sum_{i=0}^{3} P_i = [1 + M_1 + M_2 + M_3] P_0 \tag{2.101}$$

2.5 ANALYSIS OF RESULTS

2.5.1 Transient state

The effect of failure and repair rates of various sub-systems on the reliability for transient state has been examined for various choices of the probable failure rates and repair rates of some important subsystems. The system of differential equations (2.51-2.66) with initial conditions (2.67) has been solved numerically using Runge- Kutta fourth order method. Taking data for failure and repair rates of the subsystem per day. The result thus obtained are presented in tables 2.1(a-h).

The availability of the system has been calculated by varying the failure rate (α_1) of the subsystem mixture from 0.005 to 0.013 and fixing other parameters such as $\alpha_2 = 0.007$, $\alpha_3 = 0.23$, $\alpha_4 = 0.56$, $\phi = 4.2$, $\psi = 0.04$, $\mu = 2$ and $\sigma = 2.2$. The results are shown in table 2.1 (a). It shows that with increase in failure rate of mixture from 0.005 to 0.013 the availability of the system decreases approximately from 0.14% to 0.12%, whereas, it decreases approximately 3.3% to 3.2% with increase in the time from 30 days to 360 days.

The result shown in table 2.1(b) presents the availability of system computed by varying failure rate (α_2) of extruder from 0.007 to 0.023 and keeping other parameters such as $\alpha_1 = 0.005, \alpha_3 = 0.23, \alpha_4 = 0.56$, $\phi = 4.2$, $\psi = 0.04$, μ = 2 and $\sigma = 2.2$ fixed. It is noticed that with increase in failure rate (α_2) of extruder from 0.007 to 0.023, the availability of the system decreases approximately by 17.7% to 19.6%. However, it decreases approximately from 3.3% to 5.2%, with increase in the time from 30 days to 360 days.

Next, we have studied the effect of failure rate of sub system, namely die, on availability of the system by varying its failure rate (α_3) from 0.23 to 0.31. Fixing the other failure and repair rate of the subsystems $\alpha_1 = 0.005, \alpha_2 = 0.007$, $\alpha_{4=} = 0.56$, $\phi = 4.2$, $\psi = 0.04$, μ = 2 and $\sigma = 2.2$, the availability of the system is calculated using these data are shown in table 2.1(c). We say that increase in failure rate of die from 0.23 to 0.31, the availability of the system decreases approximately 0.61% to 0.58%, whereas, it decreases approximately 3.3% to 3.2% with increase in the time from 30 days to 360 days.

Taking failure and repair rates as $\alpha_1 = 0.005$, $\alpha_2 = 0.007, \alpha_3 = 0.23$, $\phi = 4.2$, $\psi = 0.04$, μ = 2 and $\sigma = 2.2$ and changing the failure of cutter from 0.56 to 0.64, the availability of the system has been calculated by using this data and results of this study are tabulated in Table 2.1(d). It indicates that with increase in failure rate of cutter from 0.56 to 0.64, the availability of the system decreases approximately 1.4% to 1.3% and it decreases also approximately 3.3% to 3.1% with the increase of time from 30 days to 360 days.

The result shown in table 2.1(e) present the availability of system computed by varying repair rate (ϕ) of mixture from 4.2 to 5.0 and fixing other parameters as $\alpha_1 = 0.005$, $\alpha_2 = 0.007, \alpha_3 = 0.23, \alpha_4 = 0.56$, ,$\psi = 0.04$,μ = 2 and $\sigma = 2.2$. It indicates with increase in repair rate of mixture (ϕ) from 4.2 to 5.0, the availability of the system increases approximately 0.01%, whereas the it decreases approximately 3.3% to 3.2% with increase in the time from 30 days to 360 days.

The results shown in the table 2.1(f) present the availability of system computed by varying repair rate (ψ) of extruder from 0.04 to 0.20 and fixing other parameters as $\alpha_1 = 0.005, \alpha_2 = 0.007, \alpha_3 = 0.23, \alpha_4 = 0.56, \phi = 4.2, \mu = 2$ and $\sigma = 2.2$. It indicates that increase in repair rate of extruder from 0.04 to 0.20 increases the availability of the system by approximately 7.0% to 10.2%, whereas the availability decreases approximately 3.3% to 0.01% with increase in the time from 30 days to 360 days.

We have next analyzed the effect of repair rate of subsystem, namely die on availability of the system by varying its repair rate (μ) of die from 2 to 6. Fixing the other parameters as $\alpha_1 = 0.005, \alpha_2 = 0.007, \alpha_3 = 0.23, \alpha_4 = 0.56, \phi = 4.2, \psi = 0.04$ and $\sigma = 2.2$. The availability of the system is calculated using these data and result are shown in table 2.1(g). It shows that increase in repair rate of die (μ) from 2 to 6 increases the availability of the system approximately 0.79% to 0.72%, whereas, the availability decreases approximately 0.3% with increase in the time from 30 days to 360 days.

The results shown in the table 2.1(h) give the availability of system computed by varying repair rate (σ) of cutter from 2.2 to 6.2 and fixing other parameters as $\alpha_1 = 0.005, \alpha_2 = 0.007, \alpha_3 = 0.23, \alpha_4 = 0.56, \phi = 4.2, \psi = 0.04$ and $\mu = 2$. It indicates that with increase in repair rate of cutter (σ) from 2.2 to 6.2, the availability of the system increases approximately 3.2% to 2.9%, whereas the availability decreases approximately 3.3% to 3.5% with increase in the time from 30 days to 360 days .

2.5.2 Steady state

We next consider the effect of various parameters on long run availability of the system. The long run availability as defined by the equation (2.101) has been computed for the various combinations of repair rates of the subsystems. It can be noticed that we have not considered all the possible combinations of the subsystems but the combinations of the prominent sub-systems have only been taken into account in this study. The following tables 2.2 (a-d) represent the long run availability of the polytube manufacturing plant.

Computation has been carried by varying the failure and repair rates of mixture as: α_1= 0, 0057, 0.0059, 0.0061, 0.0063, $\phi = 0.5, 0.7, 0.9, 1.1$ and by taking other parameters as fixed: α_2= 0.007, α_3= 0.01, α_4=0.015, ψ =2, μ =0.33, σ =0.02. The computed result as shown in table 2.1(i) reveals that increase in failure rate (α_1) of mixture decreases long run availability approximately 0.12% and the repair rate (ϕ) increses the availability approximately 0.26%.

The table 2.1(j) shows that by increasing the failure rate (α_2) of extruder, long run availability of the system is affected from 0.3% to 0.07% with the increase in failure rate. If we increase the repair rate (ψ) of extruder, the availability increases 0.6% to 0.85% with the increase in repair rate from 2.0 to 8.0.

We have next calculated the long run availability of the system by varying the failure and repair rates of die. Following values have been used for this study: α_1=0.0057, α_2 =0.007, α_4=0.015, ϕ=0.5, ψ=2.0,σ =2. Four values of α_3 (=0.01, 0.02, 0.03, 0.04) and μ (=0.33, 0.53, 0.73, 0.93) have been considered for analyzing the effect of failure and repair rates of the subsystem die. The results thus obtained have been shown in the table 2.1(k).

The failure and repair rates of all subsystems other than cutter are considered as: α_1=0.0057, α_2=0.007, α_3=0.33, ϕ=0.5, ψ =0.04, μ =0.02. The values of failure and repair rate of cutter have been taken as: α_4=0.015, 0.030, 0.045, 0.060, σ = 2.0, 4.0, 6.0, 8. 0. The long run availability has been calculated and results are shown in the table 2.1(l).

2.6 CONCLUDING OBSERVATIONS

The comparative study of the above tables 2.1(a) to 2.1(l) reveals that the sub system extruder affects the availability of the whole system more than any of the other subsystems. The effect of other subsystems on availability of polytube manufacturing plant is almost negligible. The effect of failure and repair rate of subsystem extruder on the availability of system has also been demonstrated with the help of graph (2.3- 2.4). We, thus, make an inference that management should take paramount care of sub system extruder in order to improve overall availability.

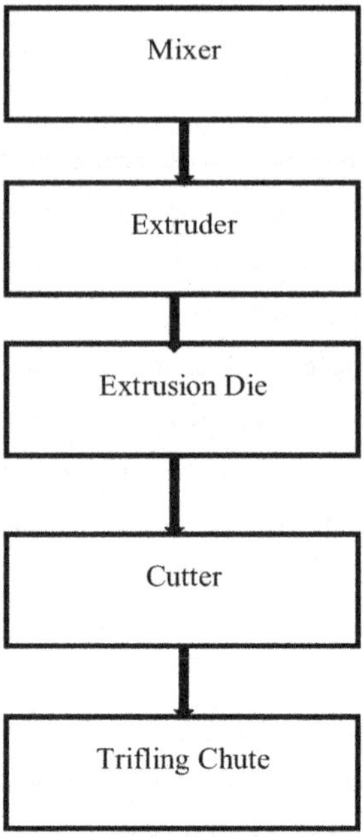

Fig. 2.1: Flow chart of polytube manufacturing plant

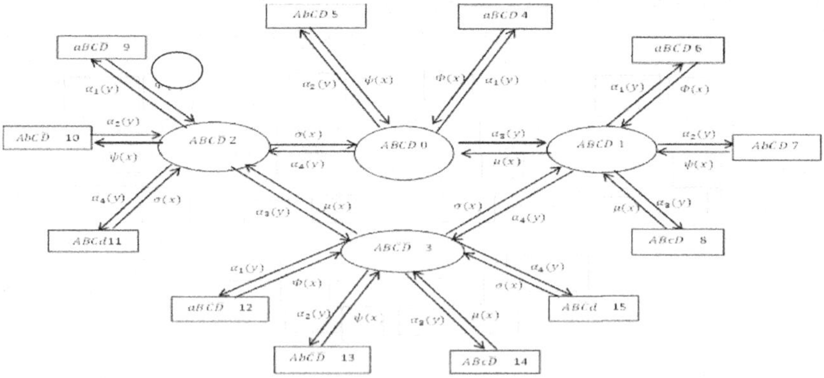

Good State: **Reduced State:** **Failed State:**

Fig. 2.2: Transition diagram of polytube manufacturing plant

Table 2.1(A): Effect of failure rate (α_2) of the subsystem Extruder (B) on availability

$\alpha_2 \rightarrow$ Time↓	.007	.011	.015	.019	.023
30	0.8394	0.7889	0.7427	0.7004	0.6616
60	0.8148	0.7559	0.7041	0.6585	0.6180
90	0.8087	0.7486	0.6965	0.6511	0.6112
120	0.8072	0.7469	0.6950	0.6498	0.6101
150	0.8068	0.7466	0.6947	0.6496	0.6099
180	0.8067	0.7465	0.6946	0.6495	0.6099
210	0.8067	0.7465	0.6946	0.6495	0.6099
240	0.8067	0.7465	0.6946	0.6495	0.6099
270	0.8067	0.7465	0.6946	0.6495	0.6099
300	0.8067	0.7465	0.6946	0.6495	0.60993
330	0.8067	0.7465	0.6946	0.6495	0.6099
360	0.8067	0.7465	0.6946	0.6495	0.6099
MTBF	285.69	263.50	245.35	232.01	220.46

Table 2.1(B): Effect of failure rate (α_3) of the subsystem Die (D) on availability

$\alpha_3 \rightarrow$ Time↓	.23	.25	.27	.29	.31
30	0.8394	0.8380	0.8364	0.8348	0.8331
60	0.8148	0.8135	0.8120	0.8054	0.8089
90	0.8087	0.8074	0.8060	0.8045	0.8029
120	0.8072	0.8059	0.8045	0.8030	0.8014
150	0.8068	0.8055	0.8041	0.8026	0.8010
180	0.8067	0.8054	0.8040	0.8025	0.8009
210	0.8067	0.8054	0.8040	0.8025	0.8009
240	0.8067	0.8054	0.8040	0.8025	0.8009
270	0.8067	0.8054	0.8040	0.8025	0.8009
300	0.8067	0.8054	0.8040	0.8025	0.8009
330	0.8067	0.8054	0.8040	0.8253	0.8009
360	0.8067	0.8054	0.8040	0.8025	0.8006
MTBF	285.69	284.13	283.55	282.56	282.47

Table 2.1(C): Effect of failure rate (α_4) of the subsystem Cutter (C) on availability

$\alpha_4 \rightarrow$ Time↓	.56	.58	.60	.62	.64
30	0.8394	0.8325	0.8301	0.8277	0.8253
60	0.8148	0.8083	0.8061	0.8038	0.8015
90	0.8087	0.8023	0.8001	0.7979	0.7956
120	0.8072	0.8008	0.7986	0.7964	0.7942
150	0.8068	0.8005	0.7983	0.7960	0.7938
180	0.8067	0.8004	0.7982	0.7960	0.7937
210	0.8067	0.8003	0.7982	0.7959	0.7937
240	0.8067	0.8003	0.7982	0.7959	0.7937
270	0.8067	0.8003	0.7982	0.7959	0.7937
300	0.8067	0.8003	0.7982	0.7959	0.7937
330	0.8067	0.8003	0.7982	0.7959	0.7937
360	0.8067	0.8003	0.7982	0.7959	0.7937
MTBF	285.69	282.45	281.58	280.83	280.75

Table 2.1(D): Effect of repair rate (ϕ) of the subsystem Mixture (A) on availability

$\phi \rightarrow$ Time↓	4.2	4.4	4.6	4.8	5.0
30	0.8394	0.8394	0.8395	0.8395	0.8395
60	0.8148	0.8148	0.8149	0.8149	0.8149
90	0.8087	0.8088	0.8088	0.8088	0.8088
120	0.8072	0.8072	0.8073	0.8073	0.8073
150	0.8068	0.8069	0.8069	0.8069	0.8070
180	0.8067	0.8068	0.8068	0.8068	0.8069
210	0.8067	0.8068	0.8068	0.8068	0.8068
240	0.8067	0.8068	0.8068	0.806	0.8068
270	0.8067	0.8068	0.8068	0.8068	0.8068
300	0.8067	0.8068	0.8068	0.8068	0.8068
330	0.8067	0.8068	0.8068	0.8068	0.8068
360	0.8067	0.8068	0.8068	0.8068	0.8068
MTBF	285.69	290.44	290.44	290.44	290.44

Table 2.1(E): Effect of repair rate (ψ) of the subsystem Extruder (B) on availability

$\psi \rightarrow$ Time↓	.04	.08	.12	.16	.20
30	0.8394	0.8733	0.8916	0.90255	0.9095
60	0.8148	0.8684	0.8906	0.9023	0.9094
90	0.8087	0.8680	0.8905	0.90231	0.9094
120	0.8072	0.8680	0.8905	0.9023	0.9094
150	0.8068	0.8680	0.8905	0.9023	0.9094
180	0.8067	0.8680	0.8905	0.9023	0.9094
210	0.8067	0.8680	0.8905	0.9023	0.9094
240	0.8067	0.8680	0.8905	0.9023	0.9094
270	0.8067	0.8680	0.8905	0.9023	0.9094
300	0.8067	0.8680	0.8905	0.9023	0.9094
330	0.8067	0.8680	0.8905	0.9023	0.9094
360	0.8067	0.8680	0.8905	0.9023	0.9094
MTBF	285.69	305.68	313.47	317.60	320.14

Table 2.1(F): Effect of repair rate (μ) of the subsystem Die (D) on availability

$\mu \rightarrow$ Time↓	2	3	4	5	6
30	0.8394	0.8440	0.8458	0.84671	0.8472
60	0.8148	0.8192	0.8208	0.8216	0.8221
90	0.8087	0.8130	0.8146	0.8155	0.8160
120	0.8072	0.8115	0.8131	0.8139	0.8144
150	0.8068	0.8111	0.8128	0.81361	0.8141
180	0.8067	0.8110	0.8127	0.8135	0.8140
210	0.8067	0.8110	0.8126	0.8135	0.8139
240	0.8067	0.8110	0.8126	0.8134	0.8139
270	0.8067	0.8110	0.8126	0.8134	0.8139
300	0.8067	0.8110	0.8126	0.8134	0.8139
330	0.8067	0.8110	0.8126	0.8134	0.8139
360	0.8067	0.8110	0.8126	0.8134	0.8139
MTBF	285.69	286.02	286.70	286.98	287.13

Table 2.1(G): Effect of repair rate (σ) of the subsystem Cutter (C) on availability

$\sigma \rightarrow$ Time↓	2.2	3.2	4.2	5.2	6.2
30	0.8394	0.8577	0.8654	0.8693	0.8715
60	0.8148	0.8320	0.8398	0.8429	0.8450

90	0.8087	0.8257	0.8328	0.8364	0.8385
120	0.8072	0.8241	0.8312	0.8348	0.8369
150	0.8068	0.8238	0.8308	0.8344	0.8365
180	0.8067	0.8237	0.8307	0.8343	0.8364
210	0.8067	0.82369	0.8307	0.8343	0.8364
240	0.8067	0.8236	0.8307	0.8343	0.8364
270	0.8067	0.8236	0.8307	0.8343	0.8364
300	0.8067	0.8236	0.8307	0.8343	0.8364
330	0.8067	0.8236	0.8307	0.8343	0.8364
360	0.8067	0.82369	0.8307	0.8343	0.8364
MTBF	285.69	290.56	293.39	294.36	295.65

Table 2.1(H): Effect of failure (α_1) and repair rates (ϕ) of the subsystem Mixture (A) on availability

$\alpha_1 \rightarrow$ $\phi \downarrow$	0.0057	0.0059	0.0061	0.0063
0.5	0.9853	0.9849	0.9845	0.9841
0.7	0.9868	0.9865	0.9860	0.9855
0.9	0.9875	0.9872	0.9860	0.9863
1.1	0.9879	0.9875	0.9872	0.9872

Table 2.1(I): Effect of failure (α_2) and repair (ψ) rates of the subsystem Extruder (B) on availability

$\alpha_2 \rightarrow$ $\psi \downarrow$	0.007	0.009	0.011	0.013
2	0.9853	0.9843	0.9832	0.9823
4	0.9869	0.9865	0.9860	0.9855
6	0.9902	0.9900	0.9898	0.9896
8	0.9913	0.9912	0.9910	0.9908

Table 2.1(J): Effect of failure (α_3) and repair (μ) rates of the subsystem Die (D) on availability

$\alpha_3 \rightarrow$ $\mu \downarrow$	0.01	0.02	0.03	0.04
0.33	0.9853	0.9853	0.9853	0.9853
0.53	0.9853	0.9853	0.9853	0.9853
0.73	0..9854	0.9854	0.9854	0.9853

0.93	0.9854	0.9854	0.9854	0.9853

Table 2.1(l): Effect of failure (α_4) and repair(σ) rates of the subsystem Cutter (C) on availability

$\alpha_4 \rightarrow$ $\sigma \downarrow$	0.015	0.030	0.045	0.060
2	0.9853	0.9853	0.9852	0.9852
4	0.9853	0.9853	0.9853	0.9852
6	0..9853	0.9852	0.9852	0.9852
8	0.9854	0.9853	0.9853	0.9852

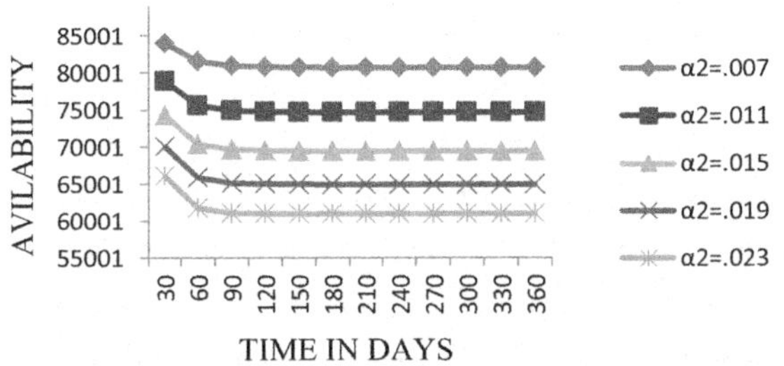

Graph 2.3: Effect of failure rate of extruder on availability

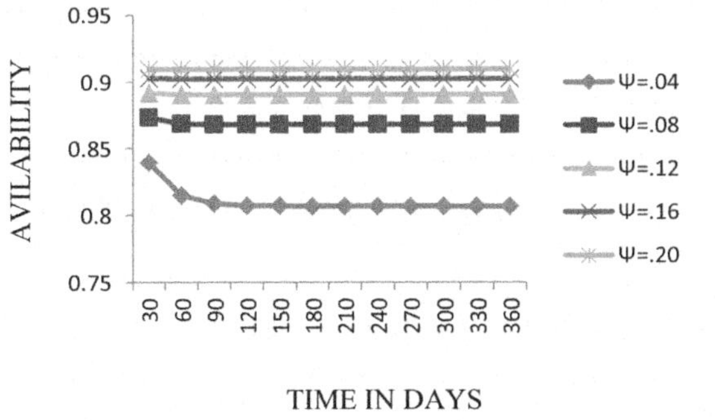

Graph 2.4: Effect of repair rate of extruder on availability

CHAPTER-III

AVAILABILITY ANALYSIS OF POLYTUBE MANUFACTURING PLANT WHEN DIE AND CUTTER FAIL INDEPENDENTELY

In the second chapter we have studied the effect of failure and repair rates on availability of the polytube manufacturing plant under the assumption that both the subsystems die and cutter fail simultaneously. Our analysis reveals that the subsystem extruder is very sensitive and it needs more care to keep the system operational. This suggestion to the concerned industry might be beneficial when subsystem die and cutter failed simultaneously. However this is not the only assumption to keep it operative. As suggested by the industry person, these two subsystems play an important role in the overall performance of this plant and fail independently also. Keeping in view the requirements of this industry, we in this chapter propose to study the availability of manufacturing plant under the assumption that the subsystems cutter and die fail independently.

The chapter is organized as follows: The polytube manufacturing plant discussed in section 2.2 of chapter II has again been chosen for the present study. In the light of new assumptions about the functioning of plant, some new notations and assumptions have been used in addition to those discussed in section 2.1. These are presented in section 3.1. Following the approach as discussed in section 2.2 a mathematical formulation of transient state has been carried out in section 3.2 for the availability analysis of polytube manufacturing plant when subsystems cutter and die are failed independently. This section also deals with steady state formulation when failure and repair rates are taken as constant. The mathematical problem developed in section 3.2 has next been solved analytically for both transient and steady state in section 3.3. The analysis of the results thus obtained is discussed in section 3.4 and conclusion drawn from the analysis is finally discussed in section 3.5.

3.1 SYSTEMS NOTATIONS AND ASSUMPTIONS

In order, to evaluate the system availability when the subsystem cutter and die fail independently, along with the assumptions and notations presented in section 2.2 of chapter II, some additional notations and assumptions are considered as given below:

The contents of this chapter have been published in **Eksploatacia Niezawodnosc- Maintenance and Reliability, 4(2011), 63-71.**

$\lambda_1(y), \lambda_2(y)$:	Transition rate of subsystem $C$and D.
$P_i(x, y, t)$: $(i = 1, \ldots, 20)$	Probability that the system is in state i at time t has an elapsed failure time y and elapsed repair time x.
$P_{4(2)}(x, y, t)$:	Probability when the sub system D is in reduced state and has an elapsed repair time x then the sub system C comes in reduced state and has an elapsed repair time x.
$P_{3(1)}(x, y, t)$:	Probability when the sub system C comes firstly in reduced state and has an elapsed repair time x then the sub system D comes in failed state and has an elapsed repair time$x$.

3.2 MATHEMATICAL FORMULATION OF THE SYSTEM WHEN TWO SUBSYSTEM CUTTER AND DIE FAIL SIMULATANEOUSELY

Following the methodology discussed in section 2.3 of chapter II and using the transition diagram (fig.2.3), the differential equations, governing the availability of the manufacturing plant when two subsystem cutter and die fail independently, has been obtained in both transition and steady state as under:

3.2.1 Transient State when failure and repair rates are variable

$$\left[\frac{d}{dt} + S_0\right] P_0(t) = M_0(t) \tag{3.1}$$

$$\left[\frac{\partial}{\partial t} + \frac{\partial}{\partial x} + \frac{\partial}{\partial y} + S_1(x, y)\right] P_1(x, y, t) = M_1(x, y, t) \tag{3.2}$$

$$\left[\frac{\partial}{\partial t} + \frac{\partial}{\partial x} + \frac{\partial}{\partial y} + S_2(x, y)\right] P_2(x, y, t) = M_2(x, y, t) \tag{3.3}$$

$$\left[\frac{\partial}{\partial t} + \frac{\partial}{\partial x} + \frac{\partial}{\partial y} + S_{3(1)}(x, y)\right] P_{3(1)}(x, y, t) = M_3(x, y, t) \tag{3.4}$$

$$\left[\frac{\partial}{\partial t} + \frac{\partial}{\partial x} + \frac{\partial}{\partial y} + S_{4(2)}(x, y)\right] P_{4(2)}(x, y, t) = M_4(x, y, t) \tag{3.5}$$

$$\left[\frac{\partial}{\partial t} + \frac{\partial}{\partial x} + \frac{\partial}{\partial y} + \phi(x)\right] P_5(x, y, t) = \alpha_1(y) P_0(t) \tag{3.6}$$

$$\left[\frac{\partial}{\partial t} + \frac{\partial}{\partial x} + \frac{\partial}{\partial y} + \psi(x)\right] P_6(x, y, t) = \alpha_2(y) P_0(t) \tag{3.7}$$

$$\left[\frac{\partial}{\partial t} + \frac{\partial}{\partial x} + \frac{\partial}{\partial y} + \phi(x)\right] P_7(x, y, t) = \alpha_1(y) P_1(x, y, t) \tag{3.8}$$

$$\left[\frac{\partial}{\partial t} + \frac{\partial}{\partial x} + \frac{\partial}{\partial y} + \psi(x)\right] P_8(x, y, t) = \alpha_2(y) P_1(x, y, t) \tag{3.9}$$

$$\left[\frac{\partial}{\partial t} + \frac{\partial}{\partial x} + \frac{\partial}{\partial y} + \mu(x)\right] P_9(x, y, t) = \alpha_3(y) P_1(x, y, t) \tag{3.10}$$

$$\left[\frac{\partial}{\partial t}+\frac{\partial}{\partial x}+\frac{\partial}{\partial y}+\phi(x)\right]P_{10}(x,y,t)=\alpha_1(y)P_2(x,y,t) \tag{3.11}$$

$$\left[\frac{\partial}{\partial t}+\frac{\partial}{\partial x}+\frac{\partial}{\partial y}+\psi(x)\right]P_{11}(x,y,t)=\alpha_2(y)P_2(x,y,t) \tag{3.12}$$

$$\left[\frac{\partial}{\partial t}+\frac{\partial}{\partial x}+\frac{\partial}{\partial y}+\sigma(x)\right]P_{12}(x,y,t)=\alpha_4(y)P_2(x,y,t) \tag{3.13}$$

$$\left[\frac{\partial}{\partial t}+\frac{\partial}{\partial x}+\frac{\partial}{\partial y}+\phi(x)\right]P_{13}(x,y,t)=\alpha_1(y)P_{4(2)}(x,y,t) \tag{3.14}$$

$$\left[\frac{\partial}{\partial t}+\frac{\partial}{\partial x}+\frac{\partial}{\partial y}+\psi(x)\right]P_{14}(x,y,t)=\alpha_2(y)P_{4(2)}(x,y,t) \tag{3.15}$$

$$\left[\frac{\partial}{\partial t}+\frac{\partial}{\partial x}+\frac{\partial}{\partial y}+\mu(x)\right]P_{15}(x,y,t)=\alpha_3(y)P_{4(2)}(x,y,t) \tag{3.16}$$

$$\left[\frac{\partial}{\partial t}+\frac{\partial}{\partial x}+\frac{\partial}{\partial y}+\sigma(x)\right]P_{16}(x,y,t)=\alpha_4(y)P_{4(2)}(x,y,t) \tag{3.17}$$

$$\left[\frac{\partial}{\partial t}+\frac{\partial}{\partial x}+\frac{\partial}{\partial y}+\phi(x)\right]P_{17}(x,y,t)=\alpha_1(y)P_{3(1)}(x,y,t) \tag{3.18}$$

$$\left[\frac{\partial}{\partial t}+\frac{\partial}{\partial x}+\frac{\partial}{\partial y}+\psi(x)\right]P_{18}(x,y,t)=\alpha_1(y)P_{3(1)}(x,y,t) \tag{3.19}$$

$$\left[\frac{\partial}{\partial t}+\frac{\partial}{\partial x}+\frac{\partial}{\partial y}+\mu(x)\right]P_{19}(x,y,t)=\alpha_1(y)P_{3(1)}(x,y,t) \tag{3.20}$$

$$\left[\frac{\partial}{\partial t}+\frac{\partial}{\partial x}+\frac{\partial}{\partial y}+\sigma(x)\right]P_{20}(x,y,t)=\alpha_1(y)P_{3(1)}(x,y,t) \tag{3.21}$$

where,

$S_0=\sum_{i=1}^{2}\alpha_i(y)+\sum_{i=1}^{2}\lambda_i$

$S_1(x,y)=\sum_{i=1}^{4}\alpha_i(y)+\mu(x)$

$S_2(x,y)=\sum_{i=1}^{4}\alpha_i+\sigma(x)$

$S_{3(1)}(x,y)=\sum_{i=1}^{4}\alpha_i(y)+\sigma(x)$

$S_{4(2)}(x,y)=\sum_{i=1}^{4}\alpha_i(y)+\mu(z)$

$$M_0(t)=\int P_5(x,y,t)\phi(x)dxdy+\int P_6(x,y,t)\psi(x)dxdy+\int P_2(x,y,t)\sigma(x)dxdy$$
$$+\int P_1(x,y,t)\mu(x)dxdy$$

$$M_1(x,y,t)=\lambda_1(y)P_0(t)+P_7(x,y,t)\phi(x)+P_8(x,y,t)\psi(x)$$
$$+P_{3(1)}(x,y,t)\sigma(x)++P_9(x,y,t)\mu(x)$$

$$M_2(x,y,t)=\lambda_2(y)P_0(t)+P_{10}(x,y,t)\phi(x)+P_{11}(x,y,t)\psi(x)+P_{12}(x,y,t)\sigma(x)$$
$$+P_{4(2)}(x,y,t)\mu(x)$$

$$M_3(x,y,t)=\alpha_4(y)P_1(x,y,t)+P_{17}(x,y,t)\phi(x)+P_{18}(x,y,t)\psi(x)+P_{19}(x,y,t)\mu(x)$$
$$+P_{20}(x,y,t)\sigma(x)$$

$$M_4(x,y,t) = \alpha_3(y)P_2(x,y,t) + P_{13}((x,y,t))\phi(x) + P_{14)}(x,y,t)\psi(x) + P_{15}(x,y,t)\mu(x)$$
$$+P_{16}(x,y,t)\sigma(x)$$

Boundary Conditions: As discussed in chapter one boundary conditions of various subsystems under the new assumptions are written as follows:

$$P_1(0,y,t) = \lambda_1(y)P_0(t) \quad (3.22)$$
$$P_2(0,y,t) = \lambda_2(y)P_0(t) \quad (3.23)$$
$$P_{3(1)}(0,y,t) = \int \alpha_4(y)P_1(x,y,t)\, dx \quad (3.24)$$
$$P_{4(2)}(0,y,t) = \int \alpha_2(y)P_2(x,y,t)dx \quad (3.25)$$
$$P_5(0,y,t) = \alpha_1(y)P_0(t) \quad (3.26)$$
$$P_6(0,y,t) = \alpha_2(y)P_0(t) \quad (3.27)$$
$$P_7(0,y,t) = \int \alpha_1(y)P_1(x,y,t)dx \quad (3.28)$$
$$P_8(0,y,t) = \int \alpha_2(y)P_1(x,y,t)\, dx \quad (3.29)$$
$$P_9(0,y,t) = \int \alpha_3(y)P_1(x,y,t)dx \quad (3.30)$$
$$P_{10}(0,y,t) = \int \alpha_1(y)P_2(x,y,t)\ \mathrm{dx} \quad (3.31)$$
$$P_{11}(0,y,t) = \int \alpha_2(y)P_2(x,y,t)\,\mathrm{dx} \quad (3.32)$$
$$P_{12}(0,y,t) = \int \alpha_1(y)P_2(x,y,t)\, dx \quad (3.33)$$
$$P_{13}(0,y,t) = \int \alpha_1(y)P_{4(2)}(x,y,t)dx \quad (3.34)$$
$$P_{14}(0,y,t) = \int \alpha_2(y)P_{4(2)}(x,y,t)\,\mathrm{dx} \quad (3.35)$$
$$P_{15}(0,y,t) = \int \alpha_3(y)P_{4(2)}(x,y,t)\, dx \quad (3.36)$$
$$P_{16}(0,y,t) = \int \alpha_4(y)P_{4(2)}(x,y,t)dx \quad (3.37)$$
$$P_{17}(0,y,t) = \int \alpha_1(y)P_{3(1)}(x,y,t)dx \quad (3.38)$$
$$P_{18}(0,y,t) = \int \alpha_2(y)P_{3(1)}(x,y,t)dx \quad (3.39)$$
$$P_{19}(0,y,t) = \int \alpha_3(y)P_{3(1)}(x,y,t)dx \quad (3.40)$$
$$P_{20}(0,y,t) = \int \alpha_4(y)P_{3(1)}(s,t)dx \quad (3.41)$$

Initial conditions: The initial conditions of the subsystems are as under:

$$P_i(x,y,0) = 0 \qquad (i = 1,\ldots,20) \quad (3.42)$$
$$P_0(0) = 1 \quad (3.43)$$

3.2.2 Steady state availability

In the process industry, management is always interested in long run availability, so we take the steady state behavior, that is $\frac{d}{dt} \to 0, \frac{\partial}{\partial t} \to 0, \frac{\partial}{\partial x} \to 0$ and $\frac{\partial}{\partial y} \to 0$ as $t \to \infty$. Then the system of differential equations (3.1-3.21) reduce to the following system of linear algebraic equations when the transition rates are constant.

$$[(\alpha_1 + \alpha_2 + \lambda_1 + \lambda_2]P_0 = \phi P_5 + \psi P_6 + \sigma P_2 + \mu P_1 \quad (3.44)$$

$$[(\alpha_1 + \alpha_2 + \alpha_3 + \alpha_4 + \mu]P_1 = \lambda_1 P_0 + \sigma P_{3(1)} + \mu P_9 + \psi P_8 + \phi P_7 \quad (3.45)$$

$$[(\alpha_1 + \alpha_2 + \alpha_3 + \alpha_4 + \sigma]P_2 = \lambda_2 P_0 + \phi P_{10} + \psi P_{11} + \sigma P_{12} + \mu P_{4(2)} \quad (3.46)$$

$$[(\alpha_1 + \alpha_2 + \alpha_3 + \alpha_4 + \sigma]P_{3(1)} = \alpha_4 P_1 + \sigma P_{20} + \mu P_{19} + \psi P_{18} + \phi P_{17} \quad (3.47)$$

$$[(\alpha_1 + \alpha_2 + \alpha_3 + \alpha_4 + \mu]P_{4(2)} = \alpha_3 P_2 + \phi P_{13} + \psi P_{14} + \sigma P_{16} + \mu P_{15} \quad (3.48)$$

$$\phi P_5 = \alpha_1 P_0 \quad (3.49)$$

$$\psi P_6 = \alpha_2 P_0 \quad (3.50)$$

$$\phi P_7 = \alpha_1 P_1 \quad (3.51)$$

$$\psi P_8 = \alpha_2 P_1 \quad (3.52)$$

$$\mu P_9 = \alpha_3 P_1 \quad (3.53)$$

$$\phi P_{10} = \alpha_1 P_1 \quad (3.54)$$

$$\psi P_{11} = \alpha_2 P_2 \quad (3.55)$$

$$\sigma P_{12} = \alpha_4 P_2 \quad (3.56)$$

$$\phi P_{13} = \alpha_1 P_{4(2)} \quad (3.57)$$

$$\psi P_{14} = \alpha_2 P_{4(2)} \quad (3.58)$$

$$\mu P_{15} = \alpha_3 P_{4(2)} \quad (3.59)$$

$$\sigma P_{16} = \alpha_4 P_{4(2)} \quad (3.60)$$

$$\phi P_{17} = \alpha_1 P_{3(1)} \quad (3.61)$$

$$\psi P_{18} = \alpha_2 P_{3(1)} \quad (3.62)$$

$$\mu P_{19} = \alpha_3 P_{3(1)} \quad (3.63)$$

$$\sigma P_{20} = \alpha_4 P_{3(1)} \quad (3.64)$$

3.3 SOLUTIONS

Equation (3.1) is a linear differential equation of first order and other equations (3.2-3.21) are partial differential equations of first order. In order to find the availability of the system, the governing equations (3.1-3.21) along with the boundary conditions (3.22-3.41) and initial conditions (3.42-3.43) have been solved analytically by using Lagrange's method and probabilities $P_i(t)(i = 1, \ldots, 20)$ for each state are obtained as follow:

$$P_0(t) = e^{-\int S_0(t)dt}\left[1 + \int M_0(t)e^{\int S_0(t)dt}dt\right] \tag{3.65}$$

$$P_1(x,y,t) = e^{-\int S_1(x,y)dx}\left[\lambda_1(y-x)P_0(t-x) + \int M_1(x,y,t)e^{\int S_1(x,y)dx}dx\right] \tag{3.66}$$

$$P_2(x,y,t) = e^{-\int S_2(x,y)dx}\left[\lambda_2(y-x)P_0(t-x) + \int M_2(x,y,t)e^{\int S_2(x,y)dx}dx\right] \tag{3.67}$$

$$P_{3(1)}(x,y,t) = e^{-\int S_{3(1)}(x,y)dx}\left[\int \alpha_4 P_1(x,y-x,t-x)dz + \int M_3(x,y,t)e^{\int S_{3(1)}(x,y)dx}dx\right] \tag{3.68}$$

$$P_{4(2)}(x,y,t) = e^{-\int S_{4(2)}(x,y)dx}\left[\int \alpha_3 P_2(x,y-x,t-x)ds + \int M_4(x,y,t)e^{\int S_{4(2)}(x,y)dx}dx\right] \tag{3.69}$$

$$P_5(x,y,t) = e^{-\int \phi(x)dx}[\alpha_1(y-x)P_0(t-x) + \int \alpha_1(y)P_0(t)\, e^{\int \phi(x)dx}dx] \tag{3.70}$$

$$P_6(x,y,t) = e^{-\int \psi(x)dx}[\alpha_2(y-x)P_0(t-y) + \int \alpha_2(y)P_0(t)\, e^{\int \psi(x)dx}] \tag{3.71}$$

$$P_7(x,y,t) = e^{-\int \phi(x)dx}[\int \alpha_1(y-x)P_1(x,y-x,t-x)dx + \int \alpha_1(y)P_1(x,y,t)e^{\int \phi(x)dx}dx \tag{3.72}$$

$$P_8(x,y,t) = e^{-\int \psi(x)dx}[\int \alpha_2(y-x)P_1(x,y-x,t-x)dx + \int \alpha_2(y)P_1(x,y,t)e^{\int \psi(x)dx}dx] \tag{3.73}$$

$$P_9(x,y,t) = e^{-\int \mu(x)dx}[\int \alpha_3(y-x)P_1(x,y-x,t-x)\, dx + \int \alpha_3(y)P_1(x,y,t)e^{\int \mu(x)dx}\, dx] \tag{3.74}$$

$$P_{10}(x,y,t) = e^{-\int \phi(x)dx}[\int \alpha_1(y-x)P_2(x,y-x,t-x)dx + \int \alpha_1(y)P_2(x,y,t)e^{\int \phi(x)dx}dx \tag{3.75}$$

$$P_{11}(x,y,t) = e^{-\int \psi(x)dx}[\int \alpha_2(y-x)P_2(x,y-x,t-x)dx + \int \alpha_2(y)P_2(x,y,t)e^{\int \psi(x)dx}dx \tag{3.76}$$

$$P_{12}(x,y,t) = e^{-\int \sigma(x)dx}[\int \alpha_4(y-x)P_2(x,y-x,t-x)dx + \int \alpha_4(y)P_2(x,y,t)e^{\int \sigma(x)dx}dx \tag{3.77}$$

$$P_{13}(x,y,t) = e^{-\int \phi(x)dx}[\int \alpha_1(y-x)P_{4(2)}(x,y-x,t-x)dx + \int \alpha_1(y)P_{4(2)}(x,y,t)e^{\int \phi(x)dx}dx] \tag{3.78}$$

$$P_{14}(x,y,t) = e^{-\int \psi(x)dx}[\int \alpha_2(y-x)P_{4(2)}(x,y-x,t-x)dx + \int \alpha_2(y)P_{4(2)}(x,y,t)e^{\int \psi(x)dx}dx]$$

(3.79)

$$P_{15}(x,y,t) = e^{-\int \mu(x)dx}[\int \alpha_3(y-x)P_{4(2)}(x,y-x,t-x)dx + \int \alpha_3(y)P_{4(2)}(x,y,t)e^{\int \mu(x)dx}dx]$$

(3.80)

$$P_{16}(x,y,t) = e^{-\int \sigma(x)dx}[\int \alpha_4(y-x)P_{4(2)}(x,y-x,t-x)dx + \int \alpha_4(y)P_{4(2)}(x,y,t)e^{\int \sigma(x)dx}d]$$

(3.81)

$$P_{17}(x,y,t) = e^{-\int \phi(x)dx}[\int \alpha_1(y-x)P_{3(1)}(x,y-x,t-x)dx + \int \alpha_1(y)P_{3(1)}(x,y,t)e^{\int \phi(x)dx}dx]$$

(3.82)

$$P_{18}(x,y,t) = e^{-\int \psi(x)dx}[\int \alpha_2(y)P_{3(1)}(x,y,t)dx + \int \alpha_2(y)P_{3(1)}(x,y,t)e^{\int \psi(x)dx}dx]$$

(3.83)

$$P_{19}(x,y,t) = e^{-\int \mu(x)dx}[\int \alpha_3(y)P_{3(1)}(x,y,t)dx + \int \alpha_3(y)P_{3(1)}(x,y,t)e^{\int \mu(x)dx}dx]$$

(3.84)

$$P_{20}(x,y,t) = e^{-\int \sigma(x)dx}[\int \alpha_4(y)P_{3(1)}(x,y,t)dx + \int \alpha_4(y)P_{3(1)}(x,y,t)\, e^{\int \sigma(x)dx}dx]$$

(3.85)

Thus all the state probabilities are obtained in terms of $P_0(t)$ as given by (3.1) and finally the time dependent availability $A(t)$ can be written as:

$$A(t) = P_0(t) + \int \sum_{i=1}^{2} P_i(x,y,t)dxdy + \int P_{4(2)}(x,y,t)dxdy + \int P_{3(1)}(x,y,t)dxdy$$

(3.86)

Next, the system of linear equations (3.44-3.64) determining the steady state availability, has been solved recursively. All the probabilities are obtained in terms of P_0 and are given as under:

$$P_1 = T_1P_0 \quad (3.87)$$

$$P_2 = T_2P_0 \quad (3.88)$$

$$P_{3(1)} = M_1P_0 \quad (3.89)$$

$$P_{4(2)} = M_2P_0 \quad (3.90)$$

where

$$M_1 = \frac{\lambda_1\alpha_4}{\sigma\mu};\ M_2 = \frac{\lambda_2\alpha_3}{\sigma\mu};\ T_1 = \frac{\lambda_1}{\mu};\ T_2 = \frac{\lambda_2}{\sigma}$$

Now using normalizing conditions $\sum_{i=0}^{20} P_i = 1$, we get

$$P_0 = \begin{bmatrix} 1 + \frac{\alpha_1}{\phi} + \frac{\alpha_2}{\psi} + \left(1 + \frac{\alpha_1}{\phi} + \frac{\alpha_2}{\psi} + \frac{\alpha_3}{\mu} + \frac{\alpha_4}{\sigma}\right) M_1 + \left(1 + \frac{\alpha_1}{\phi} + \frac{\alpha_2}{\psi} + \frac{\alpha_4}{\sigma} + \frac{\alpha_3}{\mu}\right) M_2 \\ + \left(1 + \frac{\alpha_1}{\phi} + \frac{\alpha_4}{\sigma} + \frac{\alpha_3}{\mu}\right) T_1 + \left(1 + \frac{\alpha_1}{\phi} + \frac{\alpha_2}{\psi} + \frac{\alpha_4}{\sigma}\right) T_2 \end{bmatrix}^{-1}$$

Once P_0 is known, the long run availability will be calculated as:

$$A_v = [1 + M_1 + M_2 + T_1 + T_2]P_0 \tag{3.91}$$

3.4 RESULT ANALYSIS

The steady state availability of the system represents the expected long run behavior of the system. The long run availability of the system, as defined in (3.44) has been calculated for various choices of the repair and failure rates of the sub-systems. This long run availability of the system and the result thus computed is presented in the table 3.1(a-d).

The availability of the system has been computed with the combinations of failure and repair rates of all the subsystems other than mixture of polytube manufacturing plant: $\lambda_1 = 0.001$, $\lambda_2 = 0.002$, $\alpha_2 = 0.007$, $\alpha_3 = 0.01$, $\alpha_4 = 0.015$, $\psi = 2.0$, $\mu = 0.33$, $\sigma = 0.02$. It decreases approximately 1.13% with the increase in failure rate(α_1) of mixture from 0.0057 to 0.0063 and availability increases approximately by 0.63% with the increase in repair rate (ϕ) of mixture from 0.5 to 1.1.

The result shown in table 3.1(b) represents the availability of the system for the following choices of failure and repair rates of various subsystems other than extruder : $\alpha_1 = 0.0057$, $\alpha_3 = 0.01$, $\alpha_4 = 0.015$, $\lambda_1 = 0.001$, $\lambda_2 = 0.002$, $\phi = 0.5$, $\mu = 0.33$, $\sigma = 2$, $\alpha_2 = 0.007$, 0.009, 0.011, 0.013 and four values of repair rate (ψ) = 2.0, 4.0, 6.0 , 8.0. This table shows that availability of the system decreases by approximately 2.28% with the increase in failure rate(α_2) from 0.007 to 0.013 and the availability increases by approximately 1.26% with the increase in repair rate (ψ) from 2.0 to 8.0.

The effect of failure and repair rate of die on availability has been analyzed in the table 3.1(c). Fixing failure and repair rates of all the subsystems other than die as : $\alpha_1 = 0.0057$, $\alpha_2 = 0.007$, $\alpha_4 = 0.015$, $\lambda_1 = 0.002$, $\lambda_2 = 0.002$, $\phi = 0.5$, $\psi = 2$, $\sigma = 2$.The values of failure and repair rate of die have been varied as:$\alpha_3 = 0.01, 0.02, 0.03, 0.04$ and $\mu = 0.33, 0.53, 0.73, 0.93$ respectively. The table 3.1(c) shows that increase in failure rate of die (α_3) affects long run availability by approximately 0.12% and the availability increases approximately by 0.21% with the increase in repair rate(μ) from 0.33 to 0.93.

Finally the result shown in table 3.1 (d) represents the long run availability of the system after varying the failure and repair rates of cutter. Following values have been used for this study: $\alpha_1 = 0.0057,\ \alpha_2 = 0.007,\ \alpha_3 = 0.015,\ \phi = 0.5,\ \psi = 2.0,\ \sigma\ = 2,\ \lambda_1 = 0.001, \lambda_2 = 0.002$ four values of $\alpha_4 = (0.015, 0.030, 0.045, 0.060)$ and (σ = 2.0, 4.0,6.0, 8.0) have been considered.

3.5 CONCLUDING OBSERVATIONS

Detail study of tables 3.1(a-d) shows that the subsystem extruder has maximum effect on long run availability of the system than any of other subsystems. The effect of other subsystems on availability is not so significant. In the light of this, we can suggest the management to keep utmost care of the subsystem extruder when both the subsystems die and cutter fail independently.

The comparative study of the results from the section 2.5 in chapter II with the section 3.4 in the present chapter shows that the subsystem extruder has maximum effect on the long run availability under both conditions when the subsystem cutter and die fail simultaneously and independently. There is not any appreciable change in the availability when the subsystem cutter and die fail simultaneously and independently. We can thus make an inference that management should take paramount care of the subsystem extruder in order to improve long run availability.

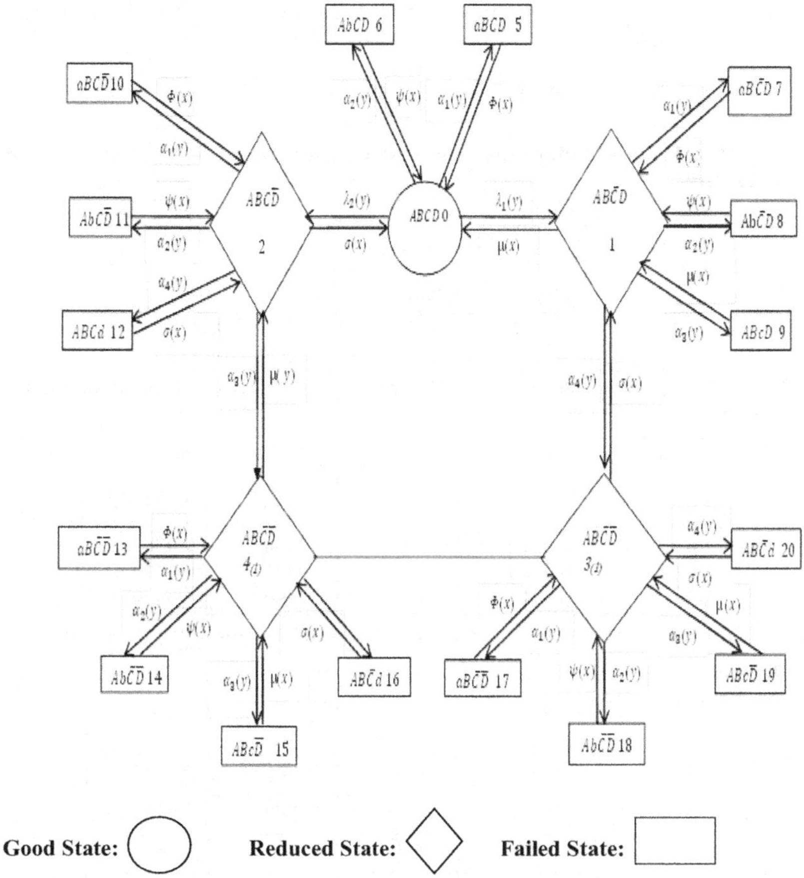

Fig. 3.1: Transition diagram of polytube manufacturing plant when Die and Cutter fail independently

Table 3.1(a): Effect of failure (α_1) and repair (ϕ) rate of the subsystem Mixture (A) on availability

$\alpha_1 \rightarrow$ $\phi \downarrow$	0.0057	0.0059	0.0061	0.0063
0.5	0.9769	0.9732	0.9694	0.9656
0.7	0.9802	0.9775	0.9748	0.9721
0.9	0.9821	0.9796	0.9770	0.9757
1.1	0.9832	0.9815	0.9797	0.9780

Table 3.1(b): Effect of failure (α_2) and repair ( ) rates of the subsystem Extruder (B) on availability

$\alpha_2 \rightarrow$ $\psi \downarrow$	0.007	0.009	0.011	0.013
2	0.9769	0.9660	0.9651	0.9541
4	0.9886	0.9881	0.9877	0.9872
6	0.9892	0.9889	0.9885	0.9882
8	0.9895	0.9892	0.9890	0.9886

Table 3.1(c): Effect of failure (α_3)and repair (μ) rates of the subsystem Die (C) on availability

$\alpha_3 \rightarrow$ $\mu \downarrow$	0.01	0.02	0.03	0.04
0.33	0.9769	0.9762	0.9760	0.9757
0.53	0.9770	0.9767	0.9464	0.9763
0.73	0.9789	0.9775	0.9772	0.9770
0.93	0.9790	0.9789	0.9780	0.9779

Table 3.1(d): Effect of failure (α_4) and repair (σ) rates of the subsystem Cutter (D) on Availability

$\alpha_4 \rightarrow$ $\sigma \downarrow$	0.5	0.030	0.045	0.060
2	0.9769	0.9763	0.9756	0.9749
4	0.9812	0.9809	0.9809	0.9802
6	0.9825	0.9823	0.9821	0.9818
8	0.9831	0.9829	0.9827	0.9826

CHAPTER-IV

PERFORMANCE ANALYSIS OF RICE MANUFACTURING PLANT

Rice is the basic food crop and being a tropical plant, flourishes comfortably in hot and humid climate. India is one of the world's largest producers of white rice, accounting for 20% of whole world production. Rice is India's preeminent crop, and is the staple food of the people of the eastern and southern parts of the country. It is one of the chief grains of India. That is why rice mills play important role for the human being.

Milling is a crucial step in post-production of rice. The basic objective of a rice milling system is to remove the husk and the bran layers, and produce edible white rice that is sufficiently milled and free of impurities. An effort has been made in this chapter to carry out the performance analysis of a rice manufacturing plant situated in Patiala, India.

This chapter is organized as follows: Section 4.1 is introductory in nature and presents the functioning of various subsystems of manufacturing plant. The detail distribution of various subsystems as well as their notation is presented in section 4.2. This section also deals with certain assumptions on which the performance of rice plant depends. As per requirement of the industry we have discussed the industry under the assumption when the rice mill works with preventive maintenance. The mathematical formulation of Chapman-Kolmogorov differential equations has been discussed in section 4.3. This section further presents the time dependent and steady state availability when failure and repair rates are variable and constant respectively. The mathematical problem thus developed has also been solved analytically and numerically for several choices of the failure and repair rates of the subsystems. The behavior analysis of the system in transient and steady states have been discussed in the section 4.4. In section 4.5 certain conclusions drawn from the present analysis are drawn.

4.1 INTRODUCTION

The rice plants are used to convert the paddy into rice. The basic objective of a rice mill system is to remove the husk and the bran layers, and produce edible white rice that is

The contents of this chapter have been published in **Communication in Computer and Information Sciences, 169 (2011), 263-274.**

sufficient milled and is made free of impurities. The figure 4.1 presents the brief process chart of the rice manufacturing plant.

4.2 SYSTEM DESCRIPTION, NOTATIONS AND ASSUMPTIONS

The rice plant consists of following six sub-systems.

Sub-System E (Elevator): Elevator (or lift) is vertical transport equipment that efficiently moves people or goods between floors (levels, decks of a building, vessel or other structures). Elevators are generally powered by electric motors that either drive traction cables and counterweight systems like a hoist, or pump hydraulic fluid to raise a cylindrical piston like a jack. These are five identical units ($E_i, i = 1,3,5,7,9$) all work in series and causes complete failure of the system if any one of the fails.

Sub-System C (Cleaning): The function of cleaner is to remove all impurities and unfilled grains from paddy. These are two identical units ($C_i, i = 1,2$) working in parallel. This unit can work with one unit in reduced capacity.

Sub-system H (Husking): This subsystem removes husk from paddy. Failure of this subsystem causes the complete failure of the system.

Sub-system S (Separation): This subsystem separates the unhusked paddy from bran rice. Failure of the subsystem causes the complete failure of the system.

Sub-system W (Whitening): The function of Whitening unit is to remove all or part of the bran layer and germ from brown rice. These are two identical units ($W_i, i = 1,2$) working in parallel. This unit can work with one subsystem in reduced capacity.

Sub-system L (Polishing): This subsystem improves the appearance of milled rice by removing the remaining bran particles and by polishing the exterior of the milled kernel. These are two identical units ($L_i, i = 10,11$).Failure of any one unit causes complete failure of the system.

The following notations will be used in the formulation of the problem:

$-$:	The Sub-system is running without any failure.
g :	The subsystem is in good state but not operative.
m :	The subsystem is under preventive maintenance.
r:	Unit is under repair.
B^z:	Indicate the working state of husking and separator machines w.r.t.

$(B = H, S)$ $(z = -, g, m, r)$.

$P_o(t)$: Probability that the system is working in full capacity.

$P_i(x, y, t)$: Probability that the system is in state i at time t and has an elapsed failure time y and elapsed repair time x, $(i = 1, \dots 5,7,9, \dots, 14)$.

$P_i(x, t)$: $(i = 6,8)$ Probability that the system is in state i at time t and has an elapsed repair time.

$E_{i\ j}^{x\ y}$: Indicates the working state of the subsystems E_i and E_j with respect to x and y, where x and y represents operative and not operative respectively. This can be written as:

$$E_{i\ j}^{x\ y} = \begin{cases} E_{3,5,7,9\ 1}^{x\ \ \ \ \ \ \ y} \\ E_{1,5,7,9\ 3}^{x\ \ \ \ \ \ \ y} \\ E_{1,3,7,9\ 5}^{x\ \ \ \ \ \ \ y} \\ E_{1,3,5,9\ 7}^{x\ \ \ \ \ \ \ y} \\ E_{1,3,5,7\ 9}^{x\ \ \ \ \ \ \ y} \end{cases},$$

where $E_{3,5,7,9\ 1}^{x\ \ \ \ \ \ \ y}$ represents the operative state of the subsystems E_3, E_5, E_7, E_9 and repair state of the subsystem E_1. Other symbol can also be represented in the similar way.

$L_{k\ \ l}^{t\ \ n}$: Indicates the working state of the subsystems L_k and L_l with respect to t and n, where t and n represents operative and under repair state. This also can be written as:

$$L_{k\ \ l}^{t\ \ n} = \begin{cases} L_{10\ \ 11}^{t\ \ \ n} \\ L_{11\ \ 10}^{t\ \ \ n} \end{cases},$$

where $L_{11\ \ 10}^{t\ \ \ n}$ represents the operative state of the subsystem D_{11} and repair state of the subsystem D_{12}.

$C_{u\ \ 3-u}^{t\ \ n}$: $t, n,$ Indicates the working state of the subsystems C_u and C_{3-u} with respect to where t and n represents operative and under repair state. This can be written as:

$$C_{u\ \ 3-u}^{t\ \ n} = \begin{cases} C_{1\ \ 2}^{t\ \ n} \\ C_{2\ \ 1}^{t\ \ n} \end{cases},$$

where $C_{1\ \ 2}^{t\ \ n}$ represents the operative state of the subsystem C_1 and repair state of the subsystem C_2

$W_v^t \;{}_{3-v}^{n}$: $t, n,$ Indicates the working state of the subsystems W_v and W_{3-v} with respect to where t and n represents operative and under repair state. This can also be written as:

$$W_v^t \;{}_{3-v}^{n} = \begin{cases} W_1^t \;{}_2^n \\ W_2^t \;{}_1^n \end{cases},$$

where $W_1^t \;{}_2^n$ represents the operative state of the subsystem W_1 and repair state of the subsystem W_2

$\lambda_i(y)$: Refers to failure rate of the subsystem $E_1, E_3, E_5, E_7, E_9, L_{10}, L_{11}, H, S$ and W from normal to failed state $(i = 1, 3, 5, 7, 9, 10, 11, 12, 13)$.

$\lambda_k(y)$: $(k = 2,4)$ Refers to failure rate of the subsystems C and W from normal to reduced state.

λ_l: $(l = 6,8)$ Refers to constant transition rate of the subsystems husk and separator which transits the sub system into the state $P_6(x,t)$ and $P_8(x,t)$ respectively after reaching to this state preventing maintenance of husking and separator machine start immediately.

b_i: $(i = 6,8)$ The probability that preventive maintenance of the subsystems husk and separator machine is carried out satisfactory and this makes the system operative.

$(1 - b_i)$: $(i = 6,8)$ The probability the preventive maintenance of H and S is carried out unsatisfactory and this makes the system to failed state thereafter.

$\mu_i(x)$: Time dependent repair rates of the subsystems $E_1, E_2, E_3, E_5, E_7, E_9, L_{10}, L_{11}$, H, S and W to return these subsystem from failed to normal state and elapsed repair time x. $(i = 1, 3, 5, 7, 9, 10, 11, 12, 13)$.

$\mu_k(x)$: $(k = 2,4)$ Time dependent repair rates of the subsystems $C$and W to return these subsystems from reduced to normal state and has an elapsed repair time x.

The assumptions, on which the present analysis is based upon, are as follows:

(i) Repair and failure rates are independent of each other and their unit is taken as per day.

(ii) Failure and Repair rates of the subsystems are taken as variable.

(iii) Performance wise, a repaired unit is as good as new one for a specified duration..

(iv) Sufficient repair facilities are provided.

(v) Service of the subsystem includes repair and/or replacement.

4.3 MATHEMATICAL MODELING OF THE SYSTEM IN TRANSIENT STATE

4.3.1 Transient state when both failure and repair rates are variable

In this section we develop the Chapman-Kolmogorov differential equation by assuming variable failure and repair rates of the subsystems by applying supplementary variable technique. In the transient state, Probability considerations give the following system of differential difference equation associated with the state transition diagram (fig. 4.2) of the system at time $(t + \Delta t)$. Using mnemonic rule, we have

$$P_0(t+\Delta t) = [1 - (\int \sum_{i=1}^{2,3,4,5,7,9,10,11,12,13} (\lambda_i(y)dy\Delta t + \lambda_6 \Delta t + \Delta t)P_0(t))$$

$$+ \int \sum_{i=1}^{2,3,4,5,7,9,10,11,12,13} \mu_i(x)\, P_i(x,y,t)dxdy\,\Delta t + \int b_6 \mu_6(x) P_6(x,t)dx\,\Delta t$$

$$+ \int b_8 \mu_8(x) P_8(x,t)dx\,\Delta t]$$

dividing both sides by Δt, we get

$$\frac{P_0(t+\Delta t) - P_0(t)}{\Delta t} = -[(\int \sum_{i=1}^{2,3,4,5,7,9,10,11,12,13} (\lambda_i(y)dy + \lambda_6 + \lambda_8)P_0(t))$$

$$+ \int \sum_{i=1}^{2,3,4,5,7,9,10,11,12,13} \mu_i(x)\, P_i(x,y,t)dxdy$$

$$+ \int b_6 \mu_6(x) P_6(x,t)dx + \int b_8 \mu_8(x) P_8(x,t)dx]$$

and on taking limit $\Delta t \to 0$, we have

$$\left[\frac{d}{dt} + \int \sum_{i=1}^{R} \lambda_i(y)dy + \lambda_6 + \lambda_8\right] P_0(t) =$$

$$+ \int \sum_{i=1}^{R} \mu_i(x)\, P_i(x,y,t)dxdy + \int b_6 \mu_6(x) P_6(x,t)dx$$

$$+ \int b_8 \mu_8(x) P_8(x,t)dx \qquad (4.1)$$

where $R = 2,3,4,5,7,9,10,11,12,13$

Similarly, for the other states, we can write the differential equation as:

$$\left[\frac{\partial}{\partial t}+\frac{\partial}{\partial x}+\frac{\partial}{\partial y}+\lambda_4(y)+\mu_2(x)\right]P_2(x,y,t)=\lambda_2(y)P_0(t)+\mu_4(x)P_{14}(x,y,t) \quad (4.2)$$

$$\left[\frac{\partial}{\partial t}+\frac{\partial}{\partial x}+\frac{\partial}{\partial y}+\lambda_2(y)+\mu_4(x)\right]P_4(x,y,t)=\lambda_4(y)P_0(t)+\mu_2(x)P_{14}(x,y,t) \quad (4.3)$$

$$\left[\frac{\partial}{\partial t}+\frac{\partial}{\partial x}+\frac{\partial}{\partial y}+\mu_i(x)\right]P_i(x,y,t)=\lambda_i(y)P_0(t) i=1,3,5,7,9,10,11. \quad (4.4)$$

$$\left[\frac{\partial}{\partial t}+\frac{\partial}{\partial x}+\frac{\partial}{\partial y}+\mu_6(x)\right]P_6(x,t)=\lambda_6P_0(t) \quad (4.5)$$

$$\left[\frac{\partial}{\partial t}+\frac{\partial}{\partial x}+\frac{\partial}{\partial y}+\mu_8(x)\right]P_8(x,t)=\lambda_8P_0(t) \quad (4.6)$$

$$\left[\frac{\partial}{\partial t}+\frac{\partial}{\partial x}+\frac{\partial}{\partial y}+\mu_{12}(x)\right]P_{12}(x,y,t)=\lambda_{12}(y)P_0(t)+(1-b_6)\mu_6(x)P_6(x,t) \quad (4.7)$$

$$\left[\frac{\partial}{\partial t}+\frac{\partial}{\partial x}+\frac{\partial}{\partial y}+\mu_{13}(x)\right]P_{13}(x,y,t)=\lambda_{13}(y)P_0(t)+(1-b_8)\mu_8(x)P_8(x,t) \quad (4.8)$$

$$\left[\frac{\partial}{\partial t}+\frac{\partial}{\partial x}+\frac{\partial}{\partial y}+\mu_4(x)+\mu_2(x)\right]P_{14}(x,y,t)=\lambda_2(y)P_4(x,y,t)+\lambda_4(y)P_2(x,y,t) \quad (4.9)$$

Boundary Conditions: As discussed in chapter one boundary conditions of various subsystems under the new assumptions are written as follows:

$$P_i(0,y,t)=\lambda_i(y)P_0(t)\ ;\ i=1,2,3,4,5,7,9,10,11,12,13 \quad (4.10)$$

$$P_i(0,t)=\int\lambda_i(y)P_0(t)dy\,;i=6,8 \quad (4.11)$$

$$P_{14}(0,y,t)=\int\lambda_2(y)P_2(x,y,t)dx+\int\lambda_4(y)P_4(x,y,t)\,dx \quad (4.12)$$

Initial Conditions: The initial conditions of the subsystems are as under:

$$P_i(x,y,0)=0\ ;\ i=1,2,3,4,5,7,9,10,11,12,13,14 \quad (4.13)$$

$$P_i(x,0)=0\ ;\ i=6,8 \quad (4.14)$$

The first order linear differential equation (4.1) and other equations (4.2-4.10) which are linear partial differential equations of first order constitute Chapman-Kolmogorov. Under the boundary conditions (4.10-4.12) and initial conditions (4.13-4.14) the equations (4.1-3.9) are solved by using Lagrange's method.

The solution thus obtained is presented as below:

$$P_0(t) = e^{-T_0 t}[1 + \int S_0(t) e^{T_0 t} dt] \quad (4.15)$$

$$P_i(x,y,t) = e^{-\int \mu_i(x)dx}[\lambda_i(y-x)P_0(t-x) + \int \lambda_i(y)P_0(t)e^{\int \mu_i(x)dx}dx] \; i = 1,3,5,7,9,10,11, \quad (4.16)$$

$$P_2(x,y,t) = e^{-\int T_1(x,y)dx}[\lambda_2(y-x)P_0(t-x) + \int S_1(x,y,t)e^{\int T_1(x,y)dx}dx] \quad (4.17)$$

$$P_4(x,y,t) = e^{-\int T_2(x,y)dx}[\lambda_4(y-x)P_0(t-x) + \int S_2(x,y,t)e^{\int T_2(x,y)dx}dx] \quad (4.18)$$

$$P_6(x,t) = e^{-\int T_3(x)dx}[\lambda_6 P_0(t-x) + \int \lambda_6 P_0(t)e^{\int T_3(x)dx}dx] \quad (4.19)$$

$$P_8(x,t) = e^{-\int T_4(x)dx}[\lambda_8 P_0(t-x) + \int \lambda_6 P_0(t)e^{\int T_4(x)dx}dx] \quad (4.20)$$

$$P_{12}(x,y,t) = e^{-\int \mu_{12}(x)dx}[\lambda_{12}(y-x)P_0(t-x) + \int S_3(x,y,t)e^{\int \mu_{12}(x)dx}dx] \quad (4.21)$$

$$P_{13}(x,y,t) = e^{-\int \mu_{13}(x)dx}[\lambda_{13}(y-x)P_0(t-x) + \int S_4(x,y,t)e^{\int \mu_{13}(x)dx}dx] \quad (4.22)$$

$$P_{14}(x,y,t) = e^{-\int T_5(x)dx}\int[\lambda_4(y-x)P_2(x,y-x,t-x)+\lambda_2(y-x)P_4(x,y-x,t-x) + S_5(x,y,t)e^{\int T_5(x)dx}]dx \quad (4.23)$$

where

$$T_0 = \int \sum_{i=1}^{2,3,4,5,7,9,10,11,12,13} \lambda_i(y)dy + \lambda_6 + \lambda_8$$

$$T_1(x,y) = \lambda_4(y)+\mu_2(x) \; ; T_2(x,y) = \lambda_2(y)+\mu_4(x); T_3(x) = (1-b_6)\mu_6(x) + b_6\mu_6(x)$$

$$T_4(x) = (1-b_8)\mu_8(x) + b_8\mu_8(x) \; ; T_5(x) = \mu_4(x)+\mu_2(x)$$

$$S_0(t) = \int \sum_{i=1}^{2,3,4,5,7,9,10,11,12,13} \mu_i(x)\, P_i(x,y,t)dxdy + \int b_6\mu_6 P_6(x,t)dx + \int b_8\mu_8 P_8(x,t)dx$$

$$S_1(x,y,t) = \lambda_2(y)P_0(t)+\mu_4(x)P_{14}(x,y,t)$$

$$S_2(x,y,t) = \lambda_4(y)P_0(t)+\mu_2(x)P_{14}(x,y,t)$$

$S_3(x,y,t) = \lambda_{12}(y)P_0(t)+(1-b_6)\mu_6(x)P_6(x,t)$

$S_4(x,y,t) = \lambda_{13}(y)P_0(t)+(1-b_8)\mu_8(x)P_8(x,t)$

$S_5(x,y,t) = \lambda_2(y)P_2(x,y,t) + \lambda_4(y)P_4(x,y,t)$

The time dependent availability $A_v(t)$ of the system is given by

$$A_v(t) = P_0(t) + \int \sum_{i=2}^{4,14} P_i(x,y,t)\,dxdy + \int \sum_{i=6}^{8} P_i(x,t)\,dx \quad (4.24)$$

4.3.2 Transient state when both failure and repair rates are constant

To find the availability of the system, when both failure and repair rates are constant, system of equitation's (4.1-4.9) reduce to more simplified form, given as:

$$\left[\frac{d}{dt} + \sum_{i=1}^{13} \lambda_i\right] P_0(t) = \sum_{i=1}^{5,7,9,10,11,12,13} \mu_i P_i(t) + b_6\mu_6 P_6(t) + b_8\mu_8 P_8(t) \quad (4.25)$$

$$\left[\frac{d}{dt} + \lambda_4 + \mu_2\right] P_2(t) = \lambda_2 P_0(t) + \mu_4 P_{14}(t) \quad (4.26)$$

$$\left[\frac{d}{dt} + \lambda_2 + \mu_4\right] P_4(t) = \lambda_4 P_0(t) + \mu_2 P_{14}(t) \quad (4.27)$$

$$\left[\frac{d}{dt} + \mu_i(x)\right] P_i(t) = \lambda_i P_0(t) \; i = 1,3,5,6,7,8,9,10,11 \quad (4.28)$$

$$\left[\frac{d}{dt} + \mu_{12}\right] P_{12}(t) = \lambda_{12} P_0(t) + (1-b_6)\mu_6 P_6(t) \quad (4.29)$$

$$\left[\frac{d}{dt} + \mu_{13}\right] P_{13}(t) = \lambda_{13} P_0(t) + (1-b_8)\mu_8 P_8(t) \quad (4.30)$$

$$\left[\frac{d}{dt} + \mu_4 + \mu_2\right] P_{14}(t) = \lambda_2 P_4(t) + \lambda_4 P_2(t) \quad (4.31)$$

Initial conditions: The initial conditions of the subsystems are as under:

$$P_i(0) = \begin{cases} 1, & i = 1 \\ 0, & \text{otherwise} \end{cases} \quad (4.32)$$

The time dependent availability $A_v(t)$ of the system can be computed as,

$$A_v(t) = \sum_{i=0,2,4,6,8,14} P_i(t) \quad (4.33)$$

4.3.3 Steady state availability

In the process industry, management is always interested in long run availability. Therefore we need steady state probabilities to calculate long run availability. This can be achieved by taking $\frac{d}{dt} \to 0$ as $t \to \infty$. Then the equations (4.25-4.31) reduce to linear system of algebraic equations when transitions rates are constant.

$$[\sum_{i=1}^{13} \lambda_i]P_0 = [\sum_{i=1}^{13} \mu_i]P_i \tag{4.34}$$

$$\mu_i P_i = \lambda_i P_0 \qquad i = 1,3,5,6,7,8,9,10,11,12,13 \tag{4.35}$$

$$[\lambda_4 + \mu_2]\, P_2 = \lambda_2 P_0 + \mu_4 P_{14} \tag{4.36}$$

$$[\lambda_2 + \mu_4] P_4 = \lambda_4 P_0 + \mu_2 P_{14} \tag{4.37}$$

$$[\mu_4 + \mu_2] P_{14} = \lambda_2 P_4 + \lambda_4 P_2 \tag{4.38}$$

Solving recursively the above system of equations (4.38-4.38), we get

$$P_2 = M_1 P_0 \tag{4.39}$$

$$P_4 = M_2 P_0 \tag{4.40}$$

$$P_{14} = M_3 P_0 \tag{4.41}$$

Now using normalizing condition

$$\sum_{i=0}^{14} P_i = 1 \tag{4.42}$$

we have

$$P_0 = \left[1 + \frac{\lambda_1}{\mu_1} + M_1 + \frac{\lambda_3}{\mu_3} + M_2 + \frac{\lambda_5}{\mu_5} + \frac{\lambda_6}{\mu_6} + \frac{\lambda_7}{\mu_7} + \frac{\lambda_8}{\mu_8} + \frac{\lambda_9}{\mu_9} + \frac{\lambda_{10}}{\mu_{10}} + \frac{\lambda_{11}}{\mu_{11}} + \frac{\lambda_{12}}{\mu_{12}} + \frac{\lambda_{13}}{\mu_{13}} + M_3\right]^{-1} \tag{4.43}$$

Once P_0 is determined, the probabilities of other states P_1, P_2, P_3,…, P_{14} can also be obtained. Finally, we can calculate the availability of the system

$$A_v = \left[1 + M_1 + M_2 + M_3 + \frac{\lambda_6}{\mu_6} + \frac{\lambda_8}{\mu_8}\right] P_0 \tag{4.44}$$

where

$$M_1 = \frac{\lambda_2}{S_4 S_5} + \frac{\mu_4 \lambda_4 \lambda_2}{S_1 S_4 S_2 S_5 S_3}$$

$$M_2 = \frac{\lambda_4}{S_2 S_3} + \frac{\mu_2 \lambda_4}{S_1 S_2 S_3} M_1$$

$$M_3 = \frac{\lambda_2}{S_1} M_2 + \frac{\lambda_4}{S_1} M_1$$

$$S_1 = \mu_4 + \mu_2$$

$$S_2 = \lambda_2 + \mu_4$$

$$S_3 = 1 - \frac{\mu_2 \lambda_2}{S_1 S_2}$$

$$S_4 = \lambda_4 + \mu_2$$

$$S_5 = 1 - \frac{\lambda_4 \mu_2}{S_1 S_2} - \frac{\mu_4 \lambda_4 \mu_2 \mu_2}{S_1 S_4 S_2 S_1 S_3}$$

4.4 ANALYSIS OF RESULT

4.4.1 Transient state

The effect of failure and repair rates of various sub-system on the reliability for transient state is examined for various values of the probable failure rates and repair rates of some important subsystems, are shown in the tables 4.1(a-h).The system of differential equations (4.25-4.31) with initial conditions (4.32) has been solved numerically using Runga- Kutte fourth order method. Failure and repair rates of the subsystem have been taken to per day.

The effect of failure rate λ_1 of the elevator on availability has been shown in the table 4.1(a) for a period 360 days divided over an interval of 30 days. It shows that in the first and 12th row, the increase in failure rate (λ_1) of elevator from .005 (once in 200 hrs.) to .010 (once in 100 hrs.) affects the availability of the system by (1.7% *to* 2.8%) whereas it affects in the 1st and 4th column (4.7% *to* 5.4%) respectively with the increase in time from(30 days to 360 *days*).

Table 4.1(b) shows the decrease in the availability in 1st and 12th row by (4.7% and 12.5%) respectively with the increase in the values of failure rate (λ_{11}) of polishing machine from 0.003 (once in 333 days) to .014 (once in 7yrs.). Further we also find in the 1st and 4th column that when the time increase from (30 days to 360 days) the availability decreases by (4.8% to 12.6%) respectively.

The 1st and 6th row of the table 4.1 (c) shows that the availability of the system affects $(0.19\% \; to \; 0.42\%)$ approximately with the increase in failure rate (λ_{12}) of husking machine (once in 19hrs. to once in 16 hrs.) whereas increase in time the availability is affected approximately $(4.7\% \; to \; 4.9\%)$ in the 1st and 4th column respectively.

The availability decreases in the 1st and 12th row by (0.18% and 0.09%) respectively with the increase in the failure rate of separating machine (λ_{13}) from(0.057% to 0.087%) in the table 4.1(d). Further we find that in the 1st and 4th column that when the time increase from (30 days to 360 days) the availability decreases by approximately (4.7% to 4.6%) respectively.

Next, in the table 4.1(e), we have calculated the effect of repair rate of elevator on availability of the system. One can see that increase in time from 30 days to 360 days decreases the availability of the system in 1st and 4th column respectively$(5.3\% \; to \; 4.4\%)$ however increase in repair rate (μ_1) of elevator from 0.029 (Once in 35hrs.) to 0.04(once in 25hrs.) availability increases approximately $(0.42\% \; to \; 1.1\%)$ in 1st and 12th row respectively.

Table 4.1(f) shows that the increase in repair rate (μ_{11}) of polishing machine from 0.011 (once in 90 hrs.) to .10(once in 10hrs.) the availability of the system increases in 1st and 12th row approximately (1.5% to 4.2%) and when the time increase from 30 days to 360days in 1st and 5th column the availability decreases approximately (5.1% to 2.2%).

The availability increases in 1st and 12th row by 0.66% to 0.55% respectively with the increase in repair rate (μ_{12}) of husking machine from $(1.10 \; to \; 4.10)$ in table 4.1 (g). Further we also find that in the 1st and 4th column that when the time increases from 30 days to 360 days the availability of the system decreases approximately $(5.1\% \; to \; 5.3\%)$ respectively.

Table 4.1(h) shows that increase in 1st and 12th row by (0.24% to 0.2%) respectively with the increase in repair rate (μ_{13}) of separating machine from(2.5 *to* 3.5). Further we also find that in the first and 4th column that by an increase in time from 30 days to 360 days the availability of the system decreases approximately (5.1% *to* 5.2%) respectively.

4.4.2 Steady State

The effect of change in various parameters on long run availability is studied in this section. The long run availability of the system is computed by alternating the failure and repair rates of the subsystem. This effect, for some important parameters, is shown in the following tables.

Table 4.2(a) shows that effect of failure and repair rate of elevator on availability of the system. It is concluded that increase in failure rate (λ_1) from (0.005 to 0.010) reduces the system availability in 1st row by (2.4%) and the when repair rate (μ_1) of elevator is increased in the 1st column from 0.029 to 0.119, the availability increases by (1.2%).

The availability decreases in the 1st row and 4th row by (15.3% and 6.6%) respectively with the increase in the failure rate λ_{11} of polishing machine from (0.003 to 0.018) as shown in the table 4.2 (b). Further we also see that in the first and 4th column that when repair rate (μ_{11}) of polishing machine is increased from (0.011 to 0.10), the availability of the system increases by (3.8% to 25.8%) respectively.

Table 4.2(c) shows that the increase in failure rate (λ_{12}) of husking machine from (0.052 to 0.82) decreases the availability of the system in the 1st row approximately (0.39%) and when repair rate μ_{12} of husking machine in the 1st column increases from (1.10 to 4.10) the availability increases approximately 0 .52%.

The availability decreases in the 1st row and 4th row approximately (0.48% and 0.24%) respectively with the increase in the failure rate λ_{13} of separating machine from(0.057 to 0.147) as shown in table 4.2 (d). Further we also see that in the first and 4th column when repair rate (μ_{13}) of separating machine is increased from (2.2 to 5.5), the availability of the system increases approximately (0.19% to 0.47%) respectively.

4.5 CONCLUDING OBSERVATIONS

The availability tables 4.1(a-h) and 4.2(a-d) provide us information about the sensitivity of the system which need more care. The result in these tables also help the management how to priorities the maintenance of subsystems so that the system will work satisfactorily for long period of time giving maximum production.

The performance analysis of rice mill help in increasing the production and quality of rice. Detailed study reveals that the polishing subsystem is critical part of the system and needs utmost care of management. Thus, the concerned managers can plan and adapt suitable maintenance practices/strategies for improving the system performance. Apart from these advantages, the system performance analysis may help to conduct cost benefit analysis, operational capability studies, inventory spare parts management and replacement decisions.

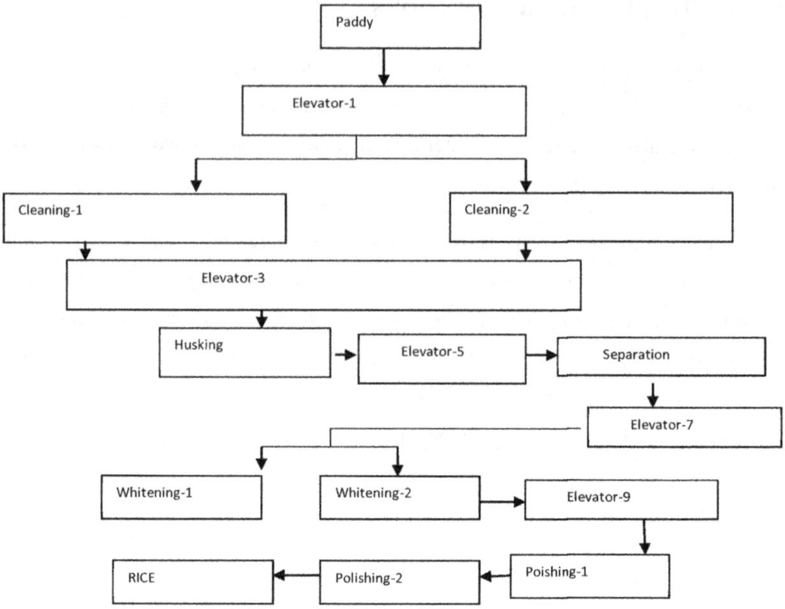

Fig. 4.1: Flow chart process diagram of rice plant.

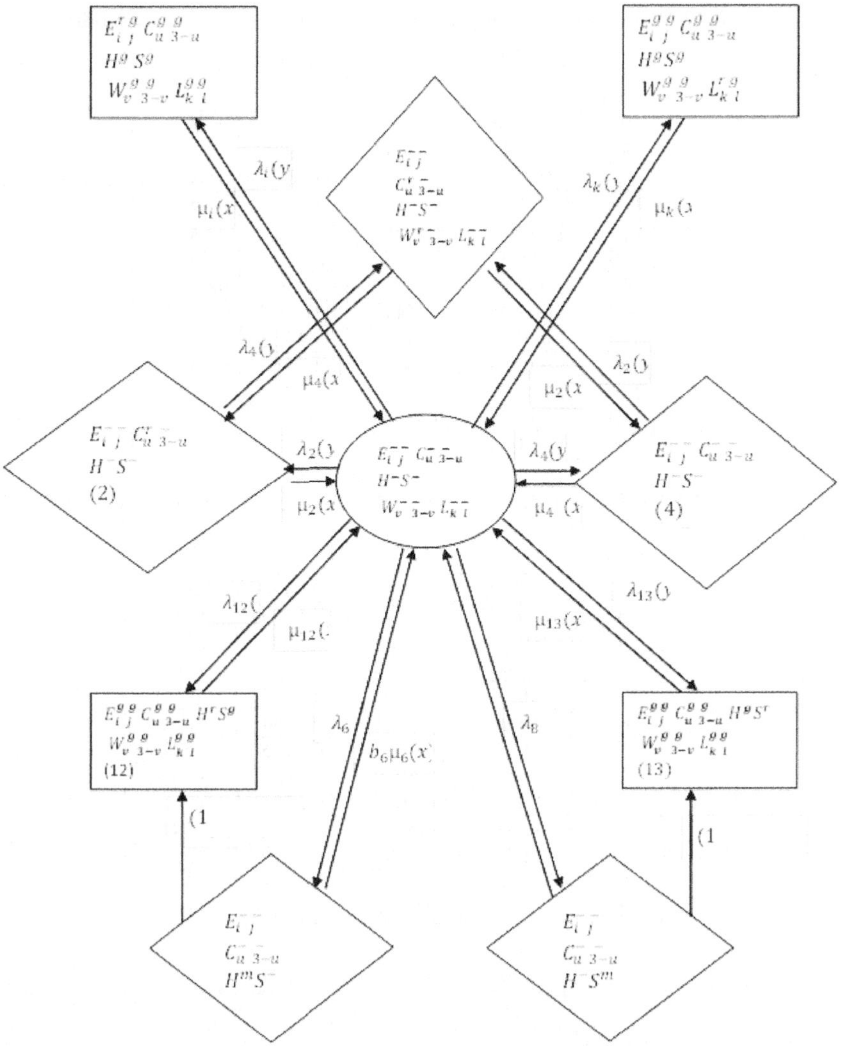

Fig. 4.2: Transition diagram of rice manufacturing plant

Table 4.1(a): Effect of failure rate (λ_1) of Elevator (E)on availability

$\lambda_1 \rightarrow$ Time↓	.005	.007	.009	.010
30	0.7038	0.6969	0.6901	0.6867
60	0.6873	0.6781	0.6690	0.6646
90	0.6777	0.6678	0.6581	0.6533
120	0.6717	0.6616	0.6518	0.6470
150	0.6677	0.6576	0.6478	0.6430
180	0.6648	0.6548	0.6450	0.6402
210	0.6627	0.6527	0.6430	0.6383
240	0.6627	0.6511	0.6414	0.6367
270	0.6596	0.6498	0.6402	0.6355
300	0.6585	0.6487	0.6391	0.6345
330	0.6575	0.6477	0.6382	0.6336
360	0.6567	0.6469	0.6374	0.6328
MTBF	234.86	231.28	229.01	226.17

Table 4.1(b): Effect of failure rate (λ_{11})of Polishing Machine (L) on availability

$\lambda_{11} \rightarrow$ Time ↓	.003	.008	.011	.014
30	0.7038	0.6814	0.6686	0.6561
60	0.6873	0.6515	0.6314	0.6121
90	0.6777	0.6334	0.6089	0.5857
120	0.6717	0.6219	0.5947	0.5693
150	0.6677	0.6143	0.5855	0.5589
180	0.6648	0.6091	0.5794	0.5527
210	0.6627	0.6055	0.5753	0.5476
240	0.6627	0.6030	0.5724	0.5454
270	0.6596	0.6011	0.5704	0.5425
300	0.6585	0.5997	0.5689	0.5410
330	0.6575	0.5985	0.5677	0.5389
360	0.6567	0.5976	0.5668	0.5310
MTBF	234.86	216.14	206.35	204.26

Table 4.1:(c) Effect of failure rate (λ_{12}) of Husking Machine (H) on availability

$\lambda_{12} \rightarrow$ Time↓	.052	.055	.058	.061
30	0.7038	0.7028	0.7024	0.7019
60	0.6873	0.6864	0.6860	0.6856
90	0.6777	0.6769	0.6765	0.6761
120	0.6717	0.6710	0.6706	0.6702
150	0.6677	0.6670	0.6666	0.6662

180	0.6648	0.6642	0.6638	0.6634
210	0.6627	0.6622	0.6618	0.6612
240	0.6627	0.6606	0.6602	0.6598
270	0.6596	0.6593	0.6589	0.6585
300	0.6585	0.6573	0.6570	0.6560
330	0.6575	0.6565	0.6560	0.6540
360	0.6567	0.6562	0.6545	0.6525
MTBF	234.86	234.60	234.42	233.82

Table 4.1 :(d) Effect of failure rate (λ_{13}) of Separation Machine(S) on availability

λ_{13} → Time↓	.057	.067	.077	.087
30	0.7038	0.7026	0.7020	0.7020
60	0.6873	0.6862	0.6856	0.6856
90	0.6777	0.6767	0.6761	0.6761
120	0.6717	0.6708	0.6702	0.6702
150	0.6677	0.6669	0.6663	0.6663
180	0.6648	0.6641	0.6635	0.6635
210	0.6627	0.6620	0.6614	0.6614
240	0.6627	0.6604	0.6598	0.6598
270	0.6596	0.6591	0.6585	0.6585
300	0.6585	0.6580	0.6575	0.6575
330	0.6575	0.6591	0.6566	0.6566
360	0.6567	0.6563	0.6558	0.6558
MTBF	234.86	234.55	234.17	234.17

Table 4.1:(e) Effect of repair rate (μ_1) of Elevator Machine (E) on availability

μ_1 → Time↓	0.029	.035	0.041	0.04
30	0.6129	0.6145	0.6158	0.6171
60	0.5960	0.5975	0.6002	0.6024
90	0.5836	0.5879	0.5912	0.5934
120	0.5771	0.5818	0.5853	0.5920
150	0.5727	0.5776	0.5813	0.5896
180	0.5697	0.5746	0.5782	0.5890
210	0.5697	0.5723	0.5759	0.5785
240	0.5674	0.5706	0.5741	0.5768
270	0.5643	0.5692	0.5727	0.5753
300	0.5633	0.5680	0.5715	0.5741
330	0.5622	0.5670	0.5706	0.5731
360	0.5613	0.5661	0.5696	0.5723
MTBF	202.08	203.07	204.02	207.27

Table 4.1(f): Effect of repair rate (μ_{11}) of Polishing Machine (L) on availability

μ_{11} → Time↓	.011	.041	.071	.10
30	0.6129	0.6195	0.6225	0.6244
60	0.5960	0.6083	0.6132	0.6155
90	0.5836	0.6038	0.6093	0.6116
120	0.5771	0.6017	0.6073	0.6097
150	0.5727	0.6005	0.6061	0.6085
180	0.5697	0.5996	0.6053	0.6076
210	0.5697	0.5988	0.6045	0.6068
240	0.5674	0.5982	0.6038	0.6060
270	0.5643	0.5975	0.6030	0.6053
300	0.5633	0.5968	0.6024	0.6046
330	0.5622	0.5961	0.6017	0.6039
360	0.5613	0.5954	0.6010	0.6032
MTBF	202.08	212.54	219.81	220.40

Table 4.1:(g) Effect of repair rate (μ_{12}) of Husking Machine (H) on availability

μ_{12} → Time↓	1.10	2.10	3.10	4.10
30	0.6129	0.6172	0.6187	0.6195
60	0.5990	0.5992	0.5996	0.6004
90	0.5836	0.5875	0.5888	0.5895
120	0.5771	0.5808	0.5822	0.5829
150	0.5727	0.5764	0.5777	0.5784
180	0.5697	0.5733	0.5746	0.5753
210	0.5697	0.5711	0.5724	0.5730
240	0.5674	0.5693	0.5706	0.5713
270	0.5643	0.5679	0.5692	0.5699
300	0.5633	0.5668	0.5680	0.5687
330	0.5622	0.5658	0.5670	0.5677
360	0.5613	0.5649	0.5662	0.5668
MTBF	202.08	202.76	203.21	203.39

Table 4.1(h): Effect of repair rate of (μ_{13}) Separation Machine (S) on availability

μ_{13} → Time↓	2.5	3.5	4.5	5.5
30	0.6129	0.6142	0.6149	0.6153
60	0.5960	0.5960	0.5962	0.5964
90	0.5836	0.5847	0.5853	0.5857

120	0.5771	0.5781	0.5788	0.5791
150	0.5727	0.5738	0.5744	0.5748
180	0.5697	0.5707	0.5713	0.5717
210	0.5697	0.5685	0.5691	0.5694
240	0.5674	0.5667	0.5673	0.5677
270	0.5643	0.5654	0.5659	0.5663
300	0.5633	0.5642	0.5648	0.5651
330	0.5622	0.5632	0.5638	0.5642
360	0.5613	0.5624	0.5629	0.5633
MTBF	202.08	202.29	202.34	202.41

Table 4.2(a): Effect of failure (λ_1) and repair (μ_1) rates of Elevator (E_1) on availability

$\lambda_1 \rightarrow$ $\mu_1 \downarrow$	.005	.007	.009	.010
.029	.6644	.6545	.6448	.6401
.059	.6776	.6724	.6679	.6649
.089	.6820	.6785	.6751	.6734
.119	.6842	.6816	.6790	.6777

Table 4.2(b): Effect of failure (λ_{11}) and repair (μ_{11}) rates of Polishing Machine (L) on availability

$\lambda_{11} \rightarrow$ $\mu_{11} \downarrow$	.003	.008	.013	.018
.011	.6644	.6038	.5533	.5105
.041	.6951	.6760	.6580	.6409
.071	.7001	.6888	.6779	.6673
.101	.7022	.6941	.6863	.7686

Table 4.2(c): Effect of failure (λ_{12}) and repair (μ_{12}) rate of Husking Machine (H) on availability

$\lambda_{12} \rightarrow$ $\mu_{12} \downarrow$	.052	.062	.072	.082
1.10	.6644	.6631	.6618	.6605
2.10	.6678	.6671	.6664	.6657
3.10	.6690	.6685	.6680	.6675
4.10	.6696	.6692	.6688	.6685

Table 4.2(d): Effect of failure (λ_{13})and repair (μ_{13}) rates of Separating Machine (*S*) on availability

λ_{13} → μ_{13} ↓	.057	.087	.117	.147
2.2	.6644	.6627	.6609	.6592
3.5	.6654	.6641	.6629	.6616
4.5	.6659	.6650	.6647	.6630
5.5	.6663	.6655	.6647	.6639

CHAPTER- V

CONCLUDING OBSERVATION

In this chapter we critically review, in brief, the work presented in the earlier chapters of thesis. Besides discussing the limitations and scope of the techniques developed in the earlier chapters for calculating the time dependent and steady state availability, the industrial significance of the results obtained in these chapters has also been discussed in this concluding chapter.

We critically review the analytical methods developed in this thesis for analyzing the effects of repair and failure rates on the time dependent availability in section 5.1 and the limitations and scope of the methods used in this thesis is also discussed in this section. In section 5.2, we discuss the industrial significance obtained in this thesis. Possible directions along which the work in this field can further proceed are finally discussed in section 5.3.

5.1 LIMITATIONS AND SCOPE OF THE METHODS PROPOSED IN THE THESIS FOR ANALYZING THE EFFECTS OF REPAIR AND FAILURE RATES ON THE TIME DEPENDENT AVAILABILITY

In the present study we have proposed the method for finding the time dependent availability of some process industrial systems. The system of partial differential equation associated with transition diagram of polytube, rice mill and steel manufacturing plants have been developed in chapter II, III, IV and V together with their respective boundary and initial conditions. The mathematical formulation has been carried out by assuming variable failure and repair rates. As management is always interested in long run availability, we have also calculated the results for long run availability.

The equations (2.1-2.16) of chapter II, (3.1-3.21) of chapter III, (4.1-4.9) of chapter IV and (5.1-5.11) of chapter V are meant for computing the time dependent availability of polytube, rice mill and steel manufacturing plant respectively. The Champan - Kolmogrorov differential difference equations thus obtained consisting of first order linear ordinary differential and partial differential equations. While obtaining the explicit formulation of these equations in each chapter, it is assumed that:

(i) Repair and failure rates are independent of each other

(ii) Failure and repair rates of the subsystems are taken as variable

(iii) Performance – wise, repair unit is as good as new one for a specific duration.

(iv) Sufficient repair facility are provided.

These equations associated with the concerned industry together with boundary and initial conditions have been solved analytically by using Lagrange's method. The analytical solutions (2.67-2.82); ((3.65-3.85); (4.15-4.23) and (5.24-5.31) thus obtained determine probabilities of various states of the manufacturing process yielding time dependent availability. The mathematical form of the probability obtained in each chapter are in such a complex form that a industry person cannot analyze the availability of manufacturing plants conveniently for any choice of the failure and repair rates of the subsystems.

The complexity of the analysis arises due to one of the assumption about the system that the failure and repair rates of the subsystems are variable which is realistic in the functioning of any process industry. Moreover, if we attempt to analyze the obtained solution numerically, a large data of failure and repair rates are required for these analytical solutions. Generally, an industry does not provide such data. Even if we take hypothetical data for these parameters, then the calculations become very tedious. However, the method is conveniently used if we take two transition states. This method may not be effective if we have more than two transition states because as large number of partial differential equations are obtained. Obviously, the analytical method discussed in this thesis is essentially suitable for the specific type of problems.

As a special case, when failure and repair rates are considered as constant, the system of partial differential equations of each chapter discussed above reduces to system of ordinary differential equations. Although, this is a very complex mathematical problem and cannot be solvable analytically but we have attempted to solve it numerically. An attempt has been made in this thesis to solve these equations numerically following Gupta et al (2005) approach. This numerical method can be conveniently applied to determine the time dependent availability of any process industry for possible choices of failure and repair rates.

In chapter II, we have calculated the availability by assuming the subsystem cutter and die fail simultaneously whereas in the chapter III we have calculated the availability when the subsystem cutter and die fail independently. In both the assumptions the subsystem extruder comes out as the sensitive subsystem. In fact, these assumptions are provided by industry itself. Therefore, a more critical analysis of this industry can further be carried out by assuming other subsystem being in failed state simultaneously and independently.

The reliability management of complex industrials system is highly sensitive issues for modeling and evaluating the performance of the system, especially during strategic maintenance planning. Rigorous efforts have been made by researchers to evolve methods to study the effect of subsystem conditions and maintenance policies on system performance. These methods involve complex computations and grow tremendously with further growth in number of subsystems.

5.2 INDUSTRIAL SIGNIFICANCE OF THE RESULTS

The methods discussed in this thesis are primarily developed for determining the time dependent availability and long run availability of the process industries such as polytube, rice and steel manufacturing plant. The performance analysis of these industry have been carried out in each chapter of the thesis using actual data of failure and repair rates of the subsystem obtained from the concerned industry. The suggestions which are coming out from analysis of our result will not only help in the maintenance but will also help for optimal planning of the resources of the respective industry management. Our suggestions to the concerned industry of manufacturing plants will also help them to improve overall performance of industry and meet their goal for maximum productivity.

5.3 SCOPE FOR FUTURE WORK

In the present analysis, the system of partial differential equations determining the time dependent availability of some process industries have been solved analytically using Lagrange's method. The forms of analytical solutions are so complicated, that one can not take any decisions using those results. How to solve such a complex mathematical problem is a challenging task for the critical analysis of the availability of the manufacturing plants. One can explore some efficient numerical methods for solving such a complex system of partial differential equations. In view of the easy availability of the computers it is not now as difficult a preposition as it appears from theoretical consideration. Keeping it in view interested researchers can develop software for analyzing the availability of the system possessing variable failure and repair rates.

In the present analysis we have analyzed the availability of manufacturing plants where certain subsystems are under preventive maintenance. However the cost incurred in maintenance has not been considered in this analysis. Also the cost incurred on repairing of the subsystems has not been taken into account while computing the long run availability of the manufacturing plants. The cost parameters are of great importance for achieving the long run availability. It will be worthwhile to extend the present analysis to study the optimal long run availability of the system considering the cost of different parameters as well. So that maximum availability of the system can be achieved by incurring minimum cost while investigating the problem of time dependent availability of the manufacturing plant the failure and repair rate of the systems are considered as variable and constant. These transition rates are assumed to follow the exponential distribution. It will be of interest to see how the present results of availability differ if we assume Weibull distribution instead of exponential distribution. In the present study we have taken the actual data from the respective industries. We are not sure on the authenticity of the data. In order to validate available data of failure and repair rates, one may explore suitable statistical test for the authenticity of the data. From the industrial view point, it will be worthwhile to extend our analysis on the risk assessment for the better performance of the industrial systems.

REFERENCES

1. Aggarwal, M., Kumar, Ashok and Garg, S. C., (1986), Stochastic behavior of a reparable system operating under fluctuating weather. Microelectronic and Reliability, vol. 26(3), pp. 555-557.

2. Agnihotri, R. K., and Satangi, S. K., (1996), Two non-identical unit system with priority based repair and inspection. Microelectronic and Reliability, vol. 36(2), pp. 279-282.

3. Agnihotri, R. K., Khare, A. and Jain, S., (2008), Reliability analysis of a system of boiler used in readymade garments industry. Journal of Reliability and Statistical Studied, vol. 1(1), pp. 33-41.

4. Attahiru Sule Alfa and T. S. S. Srinivasa Rao, (2000), Supplementary variable technique in stochastic models. Probability in Engineering and Information Sciences, vol. 14, pp. 203-218.

5. Aven, T., (1990), Availability formulae for stand by systems of similar units that are Preventive maintained. IEEE Transaction on Reliability, vol. 39 (5), pp. 603-606.

6. Barlow, R. and Hunter L., (1960), Optimum Preventive Maintenance Policies. Operation Research, vol.8, pp. 236-244.

7. Barlow, R. E. and Hunter, L. C., (1960), Reliability analysis of one unit system. Operation Research, vol.9, pp. 200-208.

8. Bellmore, M. and Jesen, P. A., (1969), An algorithm to determine the reliability of complex systems. IEEE Transaction on Reliability, R-18, pp. 169-174.

9. Chan, G. K. and Asgarpoor, S., (2001), Preventive maintenance with markov process. Proceedings of the north American Power Symposium, pp. 510-515.

10. Chander, S. and Bansal, R. K., (2005), Profit analysis of single unit reliability models at different failure modes. Proceedings international conference on reliability and Safety Engineering, IIT Kharagapur, pp. 577-588.

11. Chaung, W. K., (1987), Reliability analysis of multiple state devices. Microelectronic and Reliability, vol.27 (2), pp. 267-268.

12. Chaung, W. K., (1989), Reliability analysis of repairable and non-repairable system with common cause failure, Microelectronic and Reliability, vol. 29, pp. 545-547.

13. Chaung, W. K., (1990), A reliability analysis of K-out-of-N: G redundant system with common cause failure and critical human error. Microelectronic and Reliability, vol. 30. pp. 237-39.

14. Chery, D. H., Grogan, J. C., Holmes, W. A. and Perris, W. A., (1978), Availability analysis of chemical plant. Chemical Engineering Progress, vol. 74, pp. 55-60.

15. Chiang, D. T. and Niu, S. C., (1981), Reliability of consecutive k-out-of-n: F system. IEEE Transaction on Reliability, vol. 30, pp. 87-89.

16. Cox, D. R., (1955), Analysis of non markovian stochastic processes by the inclusion of supplementary variables, Proc. Camb. Phil. Soc., vol.55. pp. 433-441.

17. Dekker, R., (1995), On the use of operation research model for maintenance decision making. Microelectronic and Reliability, vol. 29, pp. 1321-1331.

18. Dhillon, B. S., (1977), A4 unit redundant system with common cause failures. IEEE Transaction on Reliability, vol. 2, pp. 373-374.

19. Dhillon, B. S., (1980), Mechanical component reliability under environmental stress. Microelectronic and Reliability, vol. 20, pp. 153-160.

20. Dhillon, B. S. and Singh, C., (1981), Engineering Reliability –New Techniques and Applications. John Wiley, New York.

21. Dhillon, B. S. and Misra, R. B., (1984), Reliability evaluation of system with critical human error. Microelectronic and Reliability, 124, pp. 743-759.

22. Dijkhuizen, G. V., and Heijden, M. V., (1995), Preventive maintenance and the interval availability distribution of an unreliable production system. Reliability Engineering and System Safety, vol. 66 (110), pp. 1401-1413.

23. Fukuta, J. and Kodama, M., (1974), Mission reliability for a redundant repairable system with two dissimilar units. IEEE Transaction on Reliability. vol.23 (4), pp. 280-282.

24. Gandhi, O. P. and Wani, M. F., (1999), Development of maintainability index for mechanical systems. Reliability Engineering and System Safety, vol. 65, pp. 259-270.

25. Garg, R. C., (1963), Analytical study of a complex system having two types of components. IEEE Transaction on Reliability, vol. 10, pp. 263-269.

26. Gaver, D. P., (1963) Time to failure and availability of parallel systems with repair. IEEE Transaction on Reliability, vol. 1, pp. 30-38.

27. Gaver, D. P., (1964), Failure Time for redundant repairable systems of two dissimilar elements. IEEE Transaction on Reliability, R-13, pp. 14-22.

28. Grall, A., Dieulle, I., Berenger, C. and Roussignal, M., (2002), Continuous time predictive maintenance scheduling for deteriorating system. IEEE Transaction on Reliability, vol. 51(2), pp. 141-150.

29. Gupta, P., (2003), Mathematical analysis of reliability and availability of some process industries. Ph. D. Thesis Thapar University, Patiala.

30. Gupta, P., (2004), System reliability analysis in industry. Ph. D. Thesis Kurukshetra University, Kurukshetra.

31. Gupta, P. P. and Sharma, R. K., (1987), An algebraic method for reliability calculations. Microelectronic and Reliability, vol. 27(1), pp. 19-23.

32. Gupta, P. P. and Kumar, A., (1987), Stochastic behavior of a two unit cold stand by redundant electronic equipment under human failure. Microelectronic and Reliability, vol. 27(1), pp. 15-18.

33. Gupta, P., Lal, A. K., Sharma, R. K. and Singh, J., (2005), Behavior study of cement manufacturing plant –a numerical approach. Journal of Mathematics and System Sciences, vol. 1(1), pp. 50-70.

34. Gupta, P., Lal, A. K., Sharma R. K. and Singh, J., (2005), Numerical analysis of reliability and availability of the serial processes in butter oil processing plant. International Journal of Quality and Reliability Management, vol. 22(3), pp. 303-316.

35. Gurav, S. V. and Utkin, L. V., (1995), The time dependent availability of repairable m-out- of-n and cold stand by systems by arbitrary distributions and repair facility. Microelectronic and Reliability, vol. 35 (9-10), pp. 1257-1274.

36. Gupta, R., Chaudhary, A. and Goel, R., (1993), Profit analysis of a two unit priority standby system subject to degradation and random shocks. Microelectronic and Reliability, vol. 33 (80), pp. 1073-1080.

37. Gunes, M. and Deveci, I., (2002), Reliability of service system and an application in student office. International Journal of Quality and Reliability Management, vol19 (2), pp. 206-211.

38. Gyemin Lee and Jongwoo Jeon (1999), Analysis of an N/G/1 finite queue with the supplementary variable method. Journal of Applied Mathematics and Stochastic Analysis, vol. 12(4), pp. 429-434.

39. Habchi Georges, (2002), An improved method of reliability assessment for suspended test, International Journal of Quality and Reliability Management, vol. 19 (4), pp. 454-470.

40. Hsu, L. F., (1999), Simultaneous determination of preventive maintenance and replacement policies in a Queue like Production system with minimal repair. Reliability Engineering and System Safety, vol. 63(2), pp. 161-167.

41. Jau-Chaun Ke and Yunn-Kuang Chu, (2007), Comparative analysis of availability for a redundant repairable system. Applied Mathematics and Computation, vol. 188, pp. 332-338.

42. Jayabalan, V. and Chaudhari, D., (1992), Cost optimization of maintenance scheduling for a system with assured reliability. IEEE Transaction on reliability, vol. 41 (1), pp. 21-25.

43. Kapoor, P. K. and Kapoor, K. R., (1976), Effect of standby redundancy on system reliability. IEEE Transaction on Reliability, R-25, pp. 120-121.

44. Kaushik, S. and Singh, I. P., (1994), Reliability analysis of a Naphtha Fuel oil system in a thermal power plant. Microelectronic and Reliability, vol. 34(2), pp. 362-372.

45. Kaushik, S. and Singh, I. P. (1994), Reliability analysis of feed water system in a thermal power plant. Microelectronic and Reliability, vol. 34 (4), pp.757-759.

46. Khan, M. R. R and Kabir, A. B. M. Z., (1995), Availability a simulation modeling of ammonia plant. Reliability Engineering and System Safety. vol. 48, pp. 217-227.

47. Kiureghain, A. D., and Ditlevson, O. D., (2007), Availability, reliability and downtime of system with repairable components. Reliability Engineering and system safety, vol. 92(2), pp. 66-72.

48. Kumar, D., Singh, J. and Pandey, P. C., (1989), Maintenance planning for the pulping system in paper industry. Reliability Engineering and System Safety. vol. 25 (4), pp. 293-303.

49. Kumar, D., Singh, J., and Pandey, P. C., (1990), Cost analysis of a multi-component Screening system in paper industry. Microelectronic and Reliability, vol. 30 (3), pp. 457-461.

50. Kumar, D., Singh, J. and Pandey, P. C., (1991), Behavior analysis of appear production system with different policies. Microelectronic and Reliability, vol. 31(1), pp. 47-51.

51. Kumar, D., Singh, J. and Pandey, P. C., (1992), Availability of crystallization system in the sugar industry under common cause failure. IEEE Transaction on Reliability, vol. 41(1), pp. 85-91.

52. Kumar, U. D. and Gopalan, M. N., (1997), Analysis of consecutive k-out-of-n: F systems with single repair facility. Microelectronic and Reliability, vol. 37(4), pp. 587-590.

53. Kusumdeep and Dipti (2009), Reliability Optimization of complex system through C-SOMGA. Journal of Information and Computing Sciences, vol. 4(3), pp. 163-172.

54. Kuo-HsiungWang, Wen-Li Dongand and Jyh-Bin Ke, (2006), Comparison of reliability and the availability between our systems with warm stand by components and stand by switching failures. Applied Mathematics and Computation, vol.186, pp. 1310- 1322.

55. Li, X. and Chen, J., (2004), Aging properties of the residual life length of k-out-of-n system with independent but non-identical components, Applied Stochastic Models in Business and Industry, vol. 20, pp. 143-153.

56. Li Yaun and Jian Xu, (2011), A deteriorating system with its repairman having multiple vacations. Applied Mathematics and Computation. vol. 217, pp. 4980-4989.

57. Lie, Chang Hoon and Young Ho Chun, (1986), An algorithm for preventive maintenance policy. IEEE Transaction Reliability, vol. 35 (1), pp. 71-75.

58. Ma Yau, Han, J. and Trivedi, K. S., (2001), Composite performance and availability analysis of wireless communication networks. IEEE transaction on vehicular Technology, vol. 5 (50), pp. 1216-1223.

59. Malik, S. C., Chand, P. and Singh, J., (2008), Reliability and profit evaluation of an operating system with different repair strategy subject to degradation. Journal of Mathematics and system Sciences. vol.4 (1), pp. 127-140.

60. Malhotra, M. and Trivedi, K. S., (1995), Dependability modeling using petri net based model. IEEE Transactions on Reliability, vol.44 (3), pp. 428-400.

61. Michelsen, Q., (1998), Use of reliability technology in the process industry. Reliability Engineering and System Safety, vol. 60, pp. 179-181.

62. Misra, N., Amit Kumar, Misra, and Ishwari Dutt Dhariyal, (2011), Standby redundancy allocations in series and parallel systems. Journal of Applied Probability, 48, pp. 43-55.

63. Mohan, C., Garg, R. C. and Sangal, P. P., (1962), Dependability of complex system. Operation Research, vol. 10, pp. 310-313.

64. Mokaddis, G. S. and Matta, C. H., (2010), Cost analysis of a two dissimilar-unit cold standby redundant system subject to inspection and random change in units. Journal of Mathematics and Statistics, vol. 6(3), pp. 306-31.

65. Nakagawa, T., (1980), Mean time to failure with preventive maintenance. IEEE Transactions on Reliability, vol. R-29(4), pp. 34-38.

66. Nakagawa, T., and Osaki, S., (1975), Stochastic behavior of a two unit priority stand by redundant system with repair. Microelectronic and Reliability, vol. 14, pp. 309-313.

Publisher: Eliva Press Global Ltd

Email: info@elivapress.com

Made in United States
North Haven, CT
04 November 2026

21772756R00055